A CHINESE BIOGAS MANUAL
Popularising Technology in the Countryside

Edited by
Ariane van Buren

from the original by the
Office of the Leading Group for the Propogation of Marshgas,
Sichuan (Szechuan) Province, Peoples' Republic of China

Technical Editor
Leo Pyle

Translator
Michael Cook

Practical
ACTION
PUBLISHING

Practical Action Publishing Ltd
27a Albert Street, Rugby, CV21 2SG, Warwickshire, UK
www.practicalactionpublishing.org

© Intermediate Technology Publications, 1976

First Published 1976/Digitised 2013

ISBN 10: 0 9 0303 165 5
ISBN 13: 9780903031653
ISBN Library EBook: 9781780441825
Book DOI: http://dx.doi.org/10.3362/9781780441825

Van Buren, A., (1976) *A Chinese Biogas Manual: Popularising Technology in the Countryside*,
Rugby, UK: Practical Action Publishing, <http://dx.doi.org/10.3362/9781780441825>

Since 1974, Practical Action Publishing has published and disseminated books and
information in support of international development work throughout the world. Practical
Action Publishing is a trading name of Practical Action Publishing Ltd (Company Reg.
No. 1159018), the wholly owned publishing company of Practical Action. Practical Action
Publishing trades only in support of its parent charity objectives and any profits are
covenanted back to Practical Action (Charity Reg. No. 247257, Group VAT Registration No.
880 9924 76).

Contents

List of figures

List of tables

List of plates

(Unless otherwise credited, all photographs by A. van Buren)

Acknowledgements

Bringing this manual from Chinese to English and supplementing it with material relevant to other parts of the world was only possible with the assistance of dedicated friends sustained by a sense of common purpose.

I would first like to acknowledge in a formal way the Society for Anglo-Chinese Understanding for their sponsorship of a trip which brought me to Shachiao and Xie Heng at the Chinese Embassy for her letter of interest. The China Travel Service and its excellent interpreters helped in the initial identification of the book in 1977, and gave untiring assistance and personal involvement during the trip in 1978. In particular I thank Song in Guangzhou for his indispensible role in the interview at Shachiao, and Sun for his organisation. I would like to express enormous appreciation to Li Xiaosu for his guidance in creating a framework for understanding the popularisation of technology in China, and for his patience and wit in those long hours on the train. Liang Daming merits more than I can mention here, as can be deduced from Appendix II; his generosity in sharing the experience of the commune members of Shachiao in their use of this book has given us a tangible picture of how it works in practice, given sufficient will and organisation. Finally, I have included in the publication a letter from the authors of the book with points they wanted added and three of their own photographs which they kindly supplied.

I am also grateful to Gerald Foley for his discussions in China and in London, and his standing offer of support. Dr Leo Pyle verified the technical content and proved a mainstay throughout the project. Karen Seeley struggled to get the transcription on its feet. Marion Porter, who typed out the manuscript, has done such a good job that one would think she knew Chinese. Hélène-Marie Blondel and Lin Guoyuan were indispensable in tying up loose ends, Hélène also for her initial translation in the Trans-siberian Express. Last, but foremost, I cannot thank Michael Crook enough for his fluent translation of the text, for the perspective he supplied and an unforgettable day in Peking.

The International Institute for Environment and Development (IIED) provided premises and much encouragement for the work that went into this book. I also acknowledge the support and facilities provided by the Architectural Association Graduate School. Lastly, many thanks to the Commonwealth Science Council and the Commonwealth Secretariat for their assistance in the publication of this manual, and through them, my particular appreciation to the Secretary of the

Commonwealth Science Council, Christian de Laet.

We hope that the readers of the English edition will write to us with any comments or suggestions they might have, particularly from their practical experiences, so that we may amend this manual accordingly.

A.v.B.
IIED
10 Percy Street
London W1.
February 1979

Introduction

The masses have boundless creativity. They can organise, and advance on all fronts and in all spheres where they can exert their power; they can expand and intensify production, and create for themselves daily increasing welfare enterprises.
— *Mao Zedong* (Mao Tse-Tung)

This manual has been translated virtually verbatim from the Chinese. It conveys not just the substance but the tone of Chinese technical education for rural areas. A decisive feature of China's technical education is that it encourages people to assimilate and modulate technology to their own needs — the result is that people develop themselves.

One of China's recent achievements has been the production of biogas from agricultural wastes. This practice is based upon an age-old Chinese tradition of composting human, animal and plant wastes to produce an organic fertilizer of high quality. However, by fermenting the materials in an airtight, watertight container, methane gas can be produced and collected for use as fuel for motors, cooking and lighting; and the liquid slurry can be returned to the land as fertilizer. Furthermore, digesting the wastes in a closed container kills many of the pathogens responsible for common rural diseases. What would be regarded in many countries as, at best, an efficient system of waste disposal has become in China a comprehensive, controlled method not only for improving rural health but for recycling resources and supplying energy as well.

Although the Chinese have been experimenting with biogas since the 1950s it was only in the 70s and mainly in Sichuan (Szechuan) that there was a movement to extend the practice and reproduce the equipment in a large way throughout the countryside. This manual is a result of that movement. It is a compilation of the experience gained in Sichuan, and is now being used in other provinces, as well, to guide communes which want to start their own biogas programme. Appendix II describes one such commune in Guangdong (Canton) Province and how they made use of this manual.

It has been widely asserted during the last few years that one of the main constraints on the spread of biogas technology in the rural areas of the Third World is the capital cost of the digester. As some details of the Chinese versions become known, it is becoming clear that digesters built from locally available materials can actually be built very cheaply. This monograph is the first to give full details of such digesters, and for that reason alone it should make a significant contribution. In fact, faced with the costs of digesters currently available in other countries, a number of people have concluded that work should concentrate on community rather than individual plants. The designs and experience recorded in this handbook demand a re-appraisal of that judgement.

In determining the viability of a biogas programme, it can be argued that the major factors in feasibility are more social than technical. Perhaps the most interesting and startling lessons — which are relevant not only to biogas but also to other areas of technical innovation in rural areas — can be learnt from the success recorded here that local communities have had in assimilating and adapting the technology to their own needs and conditions (Appendix II).The existence of so many different local designs contrasts sharply with experience in many other parts of the Third World, where the technical expertise has been more centralised and the diffusion has been blocked by patterns of land and property ownership. Furthermore, it is a remarkable witness to the capability of villagers and rural inhabitants to adapt a design principle to their own particular conditions.

This manual is one of the *Science in the Countryside* Editions, which are specially designed to popularise technology. It pre-supposes as little technical education as possible, relying instead on the appendix to provide basic principles; it contains minute and graphic textual explanations of how to go about drawing a plan, laying foundations, or applying mortar in paper-thin coats. As far as possible, the book is intended to short-cut the trials and errors that people in Sichuan had to go through while experimenting. It also tries to give the reader an idea of what is going on out of sight in the working digester, so that the technology becomes intelligible and so that it can be used to its full potential.

The technology is fairly sophisticated, and special care has to be taken in both construction and operation when limited to local, natural building materials. The book continually emphasises that this is not easy. In building a dome of bricks, for example, the danger of accidents is great and the construction is tricky. The advice is always to be meticulous, to try a difficult technique on a small scale before attempting it full size, and to test with care for safety and air-tightness. Every effort is made to help the reader face an unfamiliar technique. The language is simple and repetitive to inspire confidence.

The success of the biogas movement in China has depended to a large extent on their system of organisation. In rural areas there is only one hierarchy of authority. Political structure serves as a far more unifying function there than it does in Western societies. The same body of people who lead in political organisation are those who mobilise technical progress, education, and all other forms of development. In selecting leaders the criteria are rigorous. Authority is then decentralised and leaders are given full responsibility for the development and success of all projects undertaken.

There are three main levels of rural organisation: the commune (with sometimes over 50,000 people), the production brigade (which corresponds to an old village of 1,000-2,500 inhabitants), and the further subdivision of production teams with 100-200 members. The team is the basic unit in the economy of the commune and the most important level in rural organisation. Although the aim has been to evolve towards higher levels of collectivity, the team is still the fundamental unit of operation. It owns the land, distributes income, plans its own production methods, and decides how to invest its surplus revenue.

Biogas tanks are financed and built by the brigade, the team, or individual families, depending on size. The details of how this is done depends on local circumstances (see Appendix II). The general pattern, however, is that labour will be supplied by the commune members, and materials paid for from private savings in the case of small pits, or from the team's public collective fund in the case of large pits. This fund is the surplus remaining after income has been distributed to families, after a portion has been set aside for the public welfare fund (for health and education), and some sent as a contribution to the brigade fund. Collective funds are then used for investing in projects such as irrigation, chemical fertilizer, and biogas. State subsidies are sometimes available to promote a project like biogas development.

The capital cost of a digester is low — roughly 1 ¥uan per cubic metre or per person when using home-made concrete, and 5-6 ¥uan or £1.50 per cubic metre when using commercial cement. Thus, for a pit for a family of seven, the cost of materials ranges from £2.00 to £3.00. It is, however, impossible to translate these figures to other currencies. An average urban industrial worker's wage is 70 ¥uan (or £22.30) per month, but in the countryside wages are much lower and are supplemented by an allowance of grain and meat; families have private plots for vegetables and the produce can be traded or sold; also full medical care is available at negligible cost. In assessing the cost of building a biogas pit, it is the labour time which is significant. Usually, 35 working days are required — and these can extend well beyond 10 hours each — to build a seven cubic metre pit for a family of seven. This is seen not as a cost, rather as an investment, a building of technical expertise.

Before embarking on a biogas project, a brigade will send several of its members to another brigade as apprentices to learn all aspects of the technology. After assisting in construction and maintenance there, these people return as technicians to begin a biogas programme in their own brigade. They are responsible for the construction and operation of collective pits, for the training of further technicians, and for the supervision and assistance needed by individual families building their own pits.

This manual, available in bookstores throughout China, is used for background reference at the beginning of a project, and for data and guidance when adapting the methods to different geographical conditions. This is the single, standard textbook used by the Chinese when embarking on biogas projects, even when their local soil type and water table forces them to invent entirely different construction methods (see Appendix II). It has played a major role in the wide dissemination of biogas knowledge that has so far occurred, both in practical construction and in educational events such as brigade seminars Other methods of popularisation now include documentary films and television programmes.

The allocation of responsibility for maintenance, repair and operation is essential to the success of a biogas programme. Although the discipline this implies is often inconceivable outside China, the Chinese themselves see it as a rather simple affair. Every pit has a pressure gauge and each family has been carefully instructed when to let out gas. Feeding material into the pit is a continuous process with guidelines for the proportion of liquids to solids. The key to their success lies in the core of experts in the Biogas Group, whose job it is to go from team to team helping with repair and reminding people to maintain the safety standards.

The following is an excerpt from a letter from the authors of the manual:

> In the rural areas of our province, many production teams have put biogas pits under collective management, and in some teams a special group has been formed specifically for management and maintenance. The State has also trained biogas technicians, who still stay engaged in the productive activities of the commune; these can then supervise the construction, maintenance and management of biogas pits.
>
> The biogas pits built in Szechuan are of two kinds. The first type, by far the most common, is the small pit of 8-10 cubic metres capacity, built and used by an individual family. The gas produced is used for cooking and lighting. The second type is built by a production team, to have a capacity of about 100 cubic metres. The gas produced is used to power agricultural machinery, machine tools used to process agricultural and other local produce, to pump water and to generate electricity.
>
> Building pits in towns and cities is still at an experimental stage; large ones have been built by wine and spirit distilling factories as well as by waste disposal plants and human excrement treatment stations in various towns. Thus a useful product is being derived and at the same time industrial and human waste are being properly treated.
>
> Since 1974 when this book was first written, many advances have been made as a result of further scientific study, both in the techniques of gas extraction and use, for example, in the shape and design of the pit, construction techniques and materials, management and maintenance, lighting and cooking implements, the use of biogas for power in agriculture, and in the efficacy of the pit as a method of treating human and animal waste and controlling and eliminating disease. We would like the readers of the English translation to take into account that in these areas significant progress has been made.
>
> We enclose three photographs relevant to the book (Plates 3-1, 4-1, 4-3) and send our regards.
>
> *Office of the Leading Group for*
> *the Popularisation of Biogas in*
> *Sichuan Province, 26 November, 1978*

Manuals such as this are both a result and an integral part of China's system of social development. In the original text there are many passages of political rhetoric. These are part of the Chinese approach to motivating and guiding their population. They have been omitted from this edition, not because they are unimportant, but because they lose their meaning in any context other than the Chinese.

Much of what is described in this manual is relevant to other countries. It must, of course, be modified and adapted within very different geographical, social, political and economic conditions. In Bangladesh, Taiwan, Tanzania, Kenya, Ruanda, Upper Volta, Malawi and other countries, experimentation has begun. Extension services are being created and linked in India, where digesters are already operating in significant numbers.

Attempts are being made to set up an international biogas training centre for people from developing countries within the Peoples' Republic of China. Our purpose in publishing this book and our strongest recommendation to those wanting to initiate biogas development is to do so according to the particular circumstances within their country, but to learn as much as they can from countries that have already begun. Apprentices should be sent there to work with people building and maintaining pits, and come back as teachers.

1. The Advantages of Biogas for Rural Areas

Solving the fuel problem

The development of biogas is an important route to the solution of the fuel problem in the countryside, and therefore of concern to the entire rural population. The use of biogas, a fuel obtained from inexhaustible biological sources, as a replacement for solid fuels like coal and firewood, has brought about a radical change in the history of fuel for rural areas in China. It is an important technical innovation which not only solves the fuel problem for farmers and rural inhabitants, but also saves vast amounts of coal for the State. It thus plays a significant role in stimulating both industrial and agricultural production and in building up co-operation in the countryside. In Sichuan (Szechuan) Province several hundreds of thousands of commune members now have biogas. They have transformed themselves from firewood-lacking families into firewood-surplus families. In 1975, Liu Shiquan, of the 4th work-team of the 7th brigade of Donghe Commune in Zizhang County, would have had to buy 2500 kg of coal; but in October 1971 he built a biogas pit of 10 cu.m and thus solved the problem of obtaining cooking fuel for his family of nine. Not only does he not have to buy coal, but he also saves about 1000 kg of firewood. Cooking with biogas is hygenic, smokeless and far more convenient than using solid fuels, and the commune members say with joy "In the past when we cooked, the room would fill with smoke, but now it's all at the turn of a knob".

The development of biogas also solves a number of problems that were caused by the lack of fuel. Crop stalks that used to be burned for fuel can be returned to the fields as fertilizer to improve the quality of soil and permit intensive agriculture; they can also be used as fodder for pigs. The vast amount of labour that was formerly used to gather firewood and to transport coal can now be put into agricultural production. Decreasing the demand for firewood spares the forests and therefore furthers afforestation efforts. Money that would have been spent on coal and solid fuel can be saved and used to lighten the financial burden of commune members. The large amounts of coal that the State had been supplying to the countryside and the correspondingly huge expenditure on transport would also be saved and put into industrial construction. Also, after the development of biogas women are freed from heavy chores and housework and are able to participate in active production.

Stimulating agricultural production

The development of biogas is an important way to stimulate agricultural production, not only by returning crop stalks to the fields and saving labour, but also by greatly augmenting the quantity and quality of organic fertilizer. Human and animal excreta, crop stalks, and vegetable waste and leaves all become thoroughly decomposed after fermentation when sealed airtight in these biogas pits. Their nitrogen content is transformed into ammonia which is easier for plants to absorb, and therefore improves the fertilizer. According to studies by the Agricultural Institute of Sichuan Province, the ammonia content of organic fertilizer fermented for 30 days in a pit increases by 19.3% and the useful phosphatic content by 31.8%. Sealing this organic fertilizer in pits also prevents the evaporation and loss of ammonia. According to the studies of the Guangdong (Canton) Province Agricultural Institute, the ammonia content of common fertilizer that has been stored in these pits is increased by 147.2% in 30 days, whereas the same manure would actually suffer an 84.1% ammonia loss if stored in the traditional heaps and unsealed pits.

Manure that has been fermented in pits has been shown to increase agricultural yields. In experiments by the 4th work-team of the 3rd brigade of Yungxing commune in Mianyang County of Sichuan Province, the maize yield was increased by 28%, and according to an experiment by the 7th brigade of Qingdao commune in Mianzhu county, the rice yield was increased by 10%. In Pengsi county, the Agricultural Science Station of Dongfen commune reported that the wheat yield was increased by 12.5%, and according to Jinghua brigade of Chunguan commune in Gunnan County, Hubei Province, the cotton yield increased by 24.7%.

Table 1-1. *Experimental results comparing the yield of four crops when fertilised with unfermented excreta and biogas slurry (Index).*

	Maize	Rice	Cotton	Wheat
Unfermented excreta	100	100	100	100
Biogas slurry	128	110	124.7	112.5

Stalks, grass and water weeds, leaves and garbage are all good materials for producing biogas. Commune members can throw such materials into the pit at any time and thereby increase the source of fertilizer for the collective. According to our investigations, a 10 cu.m biogas pit can increase liquid fertilizer by over 200 kg per annum. In Sichuan Province the 7th work-team of the 1st brigade of Simin commune in Sindu County had, in the past, stored an annual average of just over 50 tonnes of manure. In 1975, however, every family there built a pit, and now their average store of manure has increased to 300-350 tonnes, resulting in a great rise in agricultural output. Comparing the 1974 yield of this production team with that of 1973, we find that the yield per mu (1 mu = 1/6 hectare or 660 m^2) of rape vegetable increased from 120 to 150 kg, or 25%; that of wheat from 220 to 255 kg, a 16% increase;

and rice from 305 to 350 kg, or 14.7%. So the commune members find that developing biogas provides them with ample fertilizer and a good base on which to grow crops.

Biogas as a health improvement

Developing a biogas programme is also an effective way to deal with excreta and improve the hygiene and standard of health in the countryside. One way to eliminate schistosome eggs, hookworm and other parasites is to compost all manure. Throwing all human and animal excreta into a biogas pit solves the problem of waste disposal.

The Institute of Parasitology of Sichuan Province and the Hygiene Departments of Mianzhu, Mianyang and other counties have proved many times that, after fermentation, the slurry contained, on average, over 95% *fewer* parasite eggs. In fact, in the fermentation of excreta the number of schistosome eggs and young hookworm eggs and larvae detected was reduced by 99%.

The period before the death of parasitic worm eggs is as follows:

	Summer	Winter
Schistosome eggs	14 days	37 days
Hookworm eggs	After 30 days over 90% dead	
Flat/tape worm egg	After 70 days over 99% dead	

In experiments where the environment of a biogas pit is simulated, the survival time of certain bacteria is as follows:

Dysentry bacillus (Shigella flexneri)	30 hours
Bacillus for paratyphoid	44 days

Where biogas has been developed properly, there has been effective control of parasitic diseases and schistosomiasis; the rural environment has been transformed; agricultural workers have been protected and the general standard of health has been successfully raised.*

Biogas and agricultural machinery

Developing biogas can also create a new source of fuel for the mechanization of agriculture. At present biogas is used in great quantities not only for cooking and lighting but also to drive agricultural machinery.

*For further details, see Michael G. McGarry and Jill Stainforth, eds., *Compost, Fertilizer, and Biogas Production from Human and Farm Wastes in the Peoples Republic of China*, IDRC-TS8e (International Development Research Centre, Ottawa, Canada, 1978); Richard Feacham et al, *Health Aspects of Excreta and Waste Water Management* (World Bank, Washington, D.C., 1979).

The costs involved are low, and this, of course, is welcomed by the people.

The 10th work-team of the 7th brigade of Lueping commune of Deyang county, Sichuan Province built an 81 cu.m capacity biogas pit in March 1973; by April they were able to use the biogas to operate a three horsepower internal combustion engine irrigation pump for eight to 10 hours every day and sometimes up to 14 hours. With this they irrigated over 100 mu of land, which proved a great help in fighting the drought. In Hungshun commune of Suining County, Sichuan Province, they used biogas to run a three hp petrol engine which generated electricity for a wired-speaker broadcasting system throughout the commune.

2. Basic Information about Biogas

What is biogas?

Biogas is a flammable gas produced by microbes when organic materials are fermented in a certain range of temperatures, moisture contents, and acidities, under airtight conditions. The chief component of biogas is methane. In ponds, marshes and manure pits where there is a high content of rotting organic materials, you can often see bubbles coming up to the surface and if you lit them you would see a blue flame. Because this sort of gas is often seen in ponds and marshes it is also known as marshgas.

Biogas is a form of biological energy that can be synthesised. In nature there are many raw materials from which biogas can be extracted: human and animal manure, leaves, twigs, grasses, stalks from crops, garbage, and also some agricultural and industrial wastes whose organic content is greater than 2%. These materials can produce biogas when placed out of contact with air and disintegrated by microbes. Biogas can then be used for cooking and lighting and in internal combustion engines.

Physical and chemical properties of biogas

Biogas is a mixture of methane (60-70%), carbon dioxide (CO_2), and small quantities of hydrogen sulphide (H_2S), nitrogen (N_2), hydrogen (H_2), and carbon monoxide (CO), and several other hydrocarbon compounds.

Methane itself is odourless, colourless and tasteless, but the other gases contained in biogas give it a slight smell of garlic or rotten eggs.

The weight of methane is roughly half that of air:

$$\frac{1 \text{ cu.m of methane}}{1 \text{ cu.m of air}} = \frac{0.716 \text{ kg}}{1.293 \text{ kg}} = 0.554 \text{ Kg}$$

The solubility of methane in water is very low. At $20°C$ and 1 atmosphere pressure, only three units of methane (volume) can be dissolved in 100 units of water. Methane has a chemical formula of CH_4 and a molecular weight of 16.04; it is a very stable hydrocarbon compound.

The complete combustion of methane produces a blue flame and a great amount of heat. The chemical reaction is:

$$CH_4 + 2O_2 \rightarrow CO_2 \uparrow + H_2O, \; \Delta H_c = -212 \text{ Kcal}$$

Upon complete combustion one cu.m of methane can reach a temperature of 1400°C and release 8562-9500 kilocalorie heat (1 kcal of heat will raise the temperature of 1 kg of water by 1°C). On complete combustion 1 cu.m of biogas can release 5500-6500 kcal of heat.

Methane is an important raw material for chemical industries; it can be used in the production of monochloromethane, dichloromethane, chloroform (which is also a chief source of carbon tetrachloride), acetylene, methanol etc.

Uses of biogas
Biogas can be used as a high quality fuel for cooking and lighting. One cu.m can keep one biogas lamp of a luminosity equivalent to a 60 watt electric light burning for six to seven hours. Biogas is also a superior fuel for producing power. One cu.m can keep a 1 hp internal combustion engine working for two hours — roughly equivalent to 0.6-0.7 kg of petrol. It can generate 1.25 kwh of electricity.

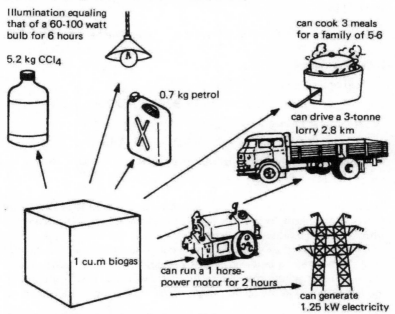

Illumination equaling that of a 60-100 watt bulb for 6 hours

5.2 kg CCl_4

0.7 kg petrol

can cook 3 meals for a family of 5-6

can drive a 3-tonne lorry 2.8 km

1 cu.m biogas

can run a 1 horse-power motor for 2 hours

can generate 1.25 kW electricity

Fig. 2-1. The uses and equivalents of biogas.

The theory and conditions for manufacturing biogas
In order to ensure the controlled fermentation of biogas and a rapid rate of production, one must understand the fundamentals of fermentation, and the conditions necessary for it to occur.

Basic theory
The fermentation process is a complicated one involving two main stages:

Stage 1: Bacteria break down complex organic materials, such as carbohydrate and chain molecules, fruit acid material, protein and fats. The disintegration produces acetic acid, lactic acid, propanoic acid, butanoic acid, methanol, ethanol and butanol, as well as carbon dioxide, hydrogen, H_2S, and other non-organic materials.

In this stage the chief micro-organisms are ones that break down polymers, fats, proteins and fruit acids, and the main action is the butanoic fermentation of polymers.

Stage 2: The simple organic materials and CO_2 that have been produced are either oxidised or reduced to methane by micro-organisms, the chief ones being the methane-producing or methanogenic micro-organisms of which there are many varieties. These need a supply of nitrogen and the latter provided by the carbon are simple organic acids and alcohols which are produced by fermentation from polymers, fats and other carbohydrates. These are broken down into methane, and carbon dioxide. This stage may be represented by the following overall reaction:

$$(C_6H_{10}O_5)n + nH_2O \text{ \textit{through the action of methane bacteria}} \quad 3nCH4 + 3nCO_2 + heat$$

Individual reactions include:

i. Acid breakdown into methane.

$$2\,C_3H_7\,COOH + H_2O \rightarrow 5CH_4 + 3CO_2$$

ii. Oxidation of ethanol by CO_2 to produce methane and acetic acid.

$$2\,CH_3CH_2OH + CO_2 \rightarrow 2CH_3\,COOH + CH_4$$

iii. Reduction with hydrogen of carbon dioxide to produce methane.

$$CO_2 + 4H_2 \rightarrow CH_4 + 2H_2O$$

Thus, complex organic materials are broken down through the action of micro-organisms to produce simple organic acids, alcohols, CO_2 etc., which are then oxidised by micro-organisms to produce methane.

This is a complex biological and chemical process and a balance must be maintained between the two stages. If the first stage proceeds at a much higher rate than the second, acid will accumulate and inhibit the fermentation in the second stage, slow it down and actually stop it.

Necessary conditions for fermentation
Since this fermentation is the result of the action of many sorts of anaerobic micro-organisms, the better the living environment of these micro-organisms, the faster the production of biogas. If proper conditions cannot be maintained, biogas production will slow or cease altogether. Optimal living conditions for these micro-organisms are:

Airtightness
None of the biological activities of anaerobic micro-organisms, including their development, breeding and metabolism, requires oxygen: in fact

they are very sensitive to the presence of oxygen. The breakdown of organic materials in the presence of oxygen will produce carbon dioxide; in airless conditions it will produce methane. If the biogas pit is not sealed to ensure the absence of air, the action of the micro-organisms and the production of biogas will be inhibited and some will escape. It is therefore crucial that the biogas pit be watertight and airtight.

Suitable temperature

The temperature for fermentation in the pit will greatly affect the production of biogas. Under suitable temperature conditions the micro-organisms become more active and gas is produced at a higher rate. Methane can be produced within a fairly wide range of temperatures, depending on prevailing conditions. Three types of fermentation are possible: at high, medium and ordinary temperatures. For high temperature fermentation, the temperature should be 50-55°C; for medium, 30-35°C; and for ordinary temperature, 10-30°C. The biogas pits in the countryside in Sichuan Province all use ordinary temperature fermentation. When the ambient temperature falls below zero, the temperature inside many biogas pits remains above 10°C, so that production is still maintained. Methane micro-organisms are very sensitive to temperature changes; a sudden change exceeding 3°C will affect production, therefore one must ensure relative stability of temperature.

Necessary nutrients

There should be plentiful material for the normal growth of the micro-organisms, and they must be able to extract plentiful nutrients from the source of fermentation. The main nutrients are carbon, nitrogen and inorganic salts. A specific ratio of carbon to nitrogen must be maintained, between 20:1 and 25:1. This ratio will vary for different raw materials, and sometimes even for the same ones. The main source of nitrogen is human and animal excrement, while the polymers in crop stalks are the main source of carbon. In order to maintain a proper ratio of carbon to nitrogen, there must be proper mixing of the human and animal excrements with polymer sources. Since there are few common materials with a suitable ratio of carbon to nitrogen production will generally not go well with only one source of material.

Water content

There must be suitable water content as the micro-organisms' excretive and other metabolic processes require water. The water content should normally be around 90% of the weight of the total contents. Both too much and too little water are harmful. With too much water the rate of production per unit volume in the pit will fall, preventing optimum use of the pit. If the water content is too low, acetic acids will accumulate, inhibiting the fermentation process and hence production; also, a rather thick scum will form on the surface. The water content should differ according to the difference in raw materials for fermentation.

23

Maintaining a suitable pH* balance

The micro-organisms require a neutral or a mildly alkaline environment — a too acidic or too alkaline environment will be detrimental. A pH between 7 and 8.5 is best for fermentation and normal gas production. The pH value for a fermentation pit depends on the ratio of acidity and alkalinity and the carbon dioxide content in the pit, the determining factor being the density of the acids. For the normal process of fermentation, the concentration of volatile acid measured by acetic acid should be below 2000 parts per million; too high a concentration will greatly inhibit the action of the methanorgenic micro-organisms.

Harmful materials

The micro-organisms that make the biogas are easily affected by many harmful materials which interfere with their livelihood. Maximum allowable concentrations of such harmful materials are as follows:

Sulphate (SO_4^{--})	5,000 parts per million
Sodium chloride (NaCl)	40,000 parts per million
Copper (Cu)	100 milligrams per litre
Chromium (Cr)	200 milligrams per litre
Nickel (Ni)	200-500 milligrams per litre
Cyanide (CN^{--})	below 25 milligrams per litre
ABS (detergent compound)	20-40 parts per million
Ammonia (NH_3)	1,500-3,000 milligrams per litre
Sodium (Na)	3,500-5,500 milligrams per litre
Potassium (K)	2,500-4,500 milligrams per litre
Calcium (Ca)	2,500-4,500 milligrams per litre
Magnesium (Mg)	1,000-1,500 milligrams per litre

Polluting substances such as these can only be accommodated if under these concentrations. They must either not be present or their concentration must be diluted, for example by the addition of water.

*pH: a measure of acidity and alkalinity; and a pH value of 7 is neutral, anything lower than 7 is acidic, anything greater than 7 is alkaline.

3. Basic Principles for the Construction of a Biogas Pit

The development of biogas in rural China parallels all other development in its emphasis on economy, self-reliance and popular mobilisation. People are encouraged to work hard and build lasting, solid and practical biogas pits of simple construction, with low demand on materials, low in cost and easy to build. In the various levels of organisation in rural Sichuan in the process of developing biogas, they engaged in ideological work, putting forward models and teaching the people by setting an example (see Appendix II). They encouraged study alongside construction work and trained leaders and technicians. From small beginnings, the development of biogas grew to large proportions in a continual spread. The main criterion was that it should be developed by the commune members themselves and that State and collective support should be subsidiary. The pits should be built using local materials and methods suited to the locality. They should be used to increase the quantity and quality of fertilizer for the collective, and also to facilitate the control of excreta and the elimination of diseases. The gas should be used to reduce or eliminate the need to buy fuel for cooking and lighting. The development of biogas is seen as an integral part of China's socialist construction and rural transformation.

One of the determinants in the production of biogas is the quality of the pit. These pits must be absolutely hermetically sealed so that the whole pit is watertight and the gas sections are airtight. This requires conscientious work and a strict scientific attitude throughout the process of construction. Each and every pit built should meet the technical specifications as any slackening of attention to quality in the building of these pits will interfere with normal gas production, affect the durability of the pits, and may even require far more work to remedy defects. It might, in the end, waste labour, building materials and time. So before the pit is built there should be exhaustive study and discussion of its size, the model to be used, the location and the materials. After thorough investigation, exact plans should be drawn up. The use of gas should be matched with the scientific disposal and treatment of excreta for fertilizer. To build toilets and pigsties above biogas pits economises on land and permits direct connection of the excreta trough with the biogas pit, so that the human and animal excreta can flow into it automatically. This also saves on labour, and increases the efficiency of excreta disposal. Additionally, in winter this also serves as insulation, helping to maintain the temperature in the biogas pit and thereby ensuring normal production of gas.

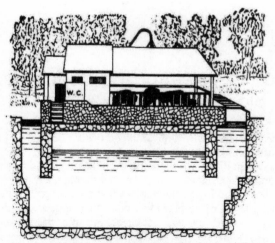

Fig. 3-1. Toilets and pigsties built above a biogas pit.

The design and construction and function of the various parts

There are many designs of biogas pits, such as those in Figures 3-2 and 3-3.

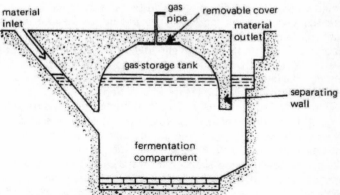

Fig. 3-2. Diagram of a circular biogas pit.

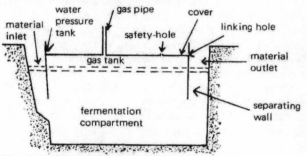

Fig. 3-3. Diagram of a rectangular biogas pit.

The material inlet

This is where the materials to be fermented enter into the fermentation compartment. This inlet should be large enough to allow easy introduction of materials. It is normally a slanting tube or trough. The lower end should open into the fermentation compartment at about mid-height. The inlet should also be linked to the excreta troughs of toilets and pigsties. The inlet should incline enough to ensure the natural flow of these materials into the fermentation compartment (see Plate 6-1).

The outlet

This is where the residue from the fermentation process is extracted. Its size should depend on the volume of the pit: there should be an adequate distance between inlet and outlet to prevent freshly incoming materials going into the outlet.

The separation wall

In the rectangular pit this separation wall creates a gas storage tank. For the round pit, the separation wall is the wall above the mouths of the inlet and outlet. The depth of the wall is normally calculated downwards from the top of the tank, so that it comes to about half the total depth of the pit (see Figures 3-2 and 3-3). If the inlet mouth is too low, the residues accumulated at the bottom of the pit may cause blockages in the inlet and outlet. Also if the separation wall is built too low it will impede air circulation, and pose a danger of suffocation for the people who clear and maintain the pit. If the separation wall is too high, it diminishes the gas storing capacity of the tank, especially at times when fertilizer is needed. If one extracts a little too much fertilizer and the liquid contents fall below the separation wall, this will cause gas to escape from the tank.

The fermentation compartment and the gas storage tank

These two sections are actually one entity. They connect the inlet and outlet and form the area where the gas is produced and stored. The middle and lower sections are the fermentation compartment, the upper is the gas storage tank, with the cover above it. When the fermentation material is let into the fermentation compartment, gas is produced through the action of micro-organisms and fermentation breakdown and it rises to the upper section and into the gas storage tank. This compartment and the tank are the principle sections of the pit, and should be strictly sealed to be completely water and airtight.

Water pressure tank

The water pressure tank is built above the gas storage tank, with the cover to the pit forming both the ceiling of the gas tank and the bottom of the water tank. Around the perimeter of the cover, a ridge about 40 cm high should be built, with a hole about 5 cm in diameter going through it just above the inlet.

As the gas rises into the gas storage tank, the liquid below it is pressed down; this raises the level of the liquid in the outlet. When it surpasses the height of the cover, the liquid flows through the hole into the water pressure tank: when the gas pressure decreases it flows back out of the water tank into the pit. As the gas is being produced, the liquid level rises; when the gas is being consumed, the liquid level falls, so that by the automatic changes in the water pressure above, the gas within the tank will be maintained at a constant pressure. Through practical experience in many regions, builders took to increasing the volume of the outlet and also increasing the height of the inlet and outlet above the cover until this served the function of a water pressure tank, which subsequently did not have to be built separately. Furthermore, this allowed them to pack in earth on top of the cover, which helped increase the pressure exerted by the cover board and also helped maintain the temperature within the pit.

The gas outlet pipe

The gas outlet pipe is set into the gas tank cover. At the bottom it opens into the gas storage tank, level with the bottom of the cover. At the upper end it may be connected to a plastic or rubber hose tubing to pipe the gas to where it will be used. The connecting pipe may be made of steel, hard plastic, or clay; it is usually 1-1.5m long depending on the amount of earth above the cover. The diameter of the pipe should be determined by the diameter of the hose that one wants to fit on to it.

Mixer

This mixer, which is not shown in the diagrams, is normally made of wooden sticks. It is used to stir the fermenting liquid and to break through the crust or scum formed on the surface of the liquid, so as to let the gas come through normally. According to accepted practice, it is not necessary to fix a mixer into small pits built for individual families. For any large pit of volume exceeding 100 cu.m a mixer should be employed so as to guarantee the normal production of gas.

Choosing an appropriate shape for the pit

At present, two main shapes of pit are being used in the countryside: round and rectangular. In regions where stone is available, it is convenient to build a circular pit out of stone slabs (flat stones having one dimension much smaller than the other two, forming thin slabs that can be used for surfacing; see Plate 3-1) or stones of irregular shape; or a rectangular pit out of stone slabs. In plains or river bed regions, rectangular and circular pits can be built out of triple concrete;* rectan-

---

*'Triple concrete' is a traditional Chinese building material, and in Sichuan it is normally made from lime, sand (which may contain small stones or pebbles), and clay, or lime, crushed clinkers and clay, mixed in specific proportions with water. Dry and liquid triple concrete differ chiefly in their water content. Dry triple concrete contains roughly 17-22% water and must be pounded into place during construction. The liquid version contains more water and is spread onto existing surfaces; its use in this province is not widespread.

On pages 39-44, two other ways of making and using triple concrete are described. They differ from the above in that plaster of Paris (gypsum) or cement is substituted for the clay.

Plate 3-1. Circular biogas pit of stone slabs during construction.

gular pits can alternatively be built with egg stones (these are flattish stones taken from river beds, roughly egg-shaped and about 50 cm in diameter.

According to geometrical principles, of all shapes the sphere has the smallest surface area for a given volume — 24% less than that of a cube whose surface area in turn is less than that of any parallelepiped of similar volume. For this reason a spherical pit is the most economical in building materials. A spherical pit built with stone slabs will require about 40% less work and material than a rectangular pit built with oblong stones and having the same volume (oblong stones have one dimension much longer than the other two and are generally used for structural purposes). A spherical shape allows a large internal volume and a small opening, which facilitates sealing to prevent air- and water-leakage. This shape also produces an even pressure on all sides, which facilitates construction and therefore makes it a good model in popular-ising the technology.

In regions with sheer rock, gravel or shale, excavation can proceed directly and there will be no need to line the pit. This saves a great deal in building materials, transport and labour. Normally the cost per cu.m is about 1 ¥uan (£0.30), a cost low enough to be popularly acceptable. This model is therefore the best, where these conditions prevail.

Converting old manure pits or vegetable stores into biogas pits is another popular and economical method. In planning the construction of a pit, one must take advantage of local materials and choose a type of construction suited to the soil conditions and the availability of building materials.

Choosing a suitable foundation for the pit
The choice of a suitable base is a key determinant in the final quality of the pit. Before construction, a careful study should be made of the soil conditions and the underground water level. One should not proceed blindly. In order to ensure that the pit will last, one should choose a site where the soil is suitably firm and the underground water level low. Wherever possible, pits should be built a fair distance away from woods, trees, and bamboo groves, so that roots will not come into the pit or cause cracks. It is also possible to cut off roots at various points where necessary and to spread some lime over the surface to stop the root growing in that direction and into the pit.

At the same time the site of the pit should be close enough to the place of use to economise on connecting hose and to reduce the resistance to gas flow in the hose, which could lead to a drop in gas pressure at the appliance.

Volume of the pit
The volume of the pit should be determined by estimating how much gas will be needed and how it will be used. Experience shows that in the countryside a family of five will require one cu.m per day for cooking and lighting. In summer each cu.m of pit will produce 0.15-0.2 cu.m per day, in winter 0.1-0.15 cu.m per day. By improving management techniques, this production can be bettered. So, when building pits one should go by the rule that 1.5-2.0 cu.m of volume should be allowed per head. One may also calculate a suitable volume for the pit by the following rule: that for two or less people there should be no more than 3 cu.m per head; for three to five people, no more than 2 cu.m per head, and for over five people there should be no more than 1.5 cu.m per head allowed. With this sizing and suitable management one should normally be able to produce enough biogas during summer and autumn for cooking and lighting, with a little left over. With lower winter temperatures and a smaller production of gas, cooking requirements will still be met.

It is one-sided to think that the larger the pit, the more the gas produced. It has been said: "The success of a pit depends on management". Though the volume of a pit may be small, with good scientific management it is possible to produce a lot of gas and guarantee supply. Even with a large pit, if raw materials are not steadily added and proper maintenance kept up, the production of gas may be less than with a small pit. Also, for larger pits, more materials and work are needed, raising the capital cost, which becomes a burden on the people. Therefore the idea that it is better to have a large pit than a small one must be overcome. Of course, neither should the volume of the pit be so small that the production of gas is insufficient and needs are not fulfilled. Production teams and production brigades, enterprises and various organisations that make use of biogas to drive machines for agricultural production and small industry or for pumping water and generating electricity, often need fairly large pits. The volume of the pit in such cases should depend on the power and type of machine to be driven, the gas consumption per day, and the source of the material for fermentation. Generally speaking, for each horse-power of an internal combustion engine, half a cu.m of gas is needed for every hour of work.

Preparation of materials and time taken for construction

Before starting work one should calculate from the volume of the pit, the rough quantities of materials needed for construction, and these should be made ready. The work should take place at slack season, in co-ordination with seasonal farm work and at the decision of the production team. In low and plains areas, especially where the water level is high, it is best to build the pit during the winter or spring. In areas where the earth freezes, work should be done before this happens, building the pit below the soil's freezing level. This not only facilitates work, but also ensures the normal production of gas in winter.

4. Different Designs of Biogas Pits

Circular Pit made of Round Stone Slabs and Irregularly Shaped Stones*
Pits built from round stone slabs and irregularly shaped stones have a large internal volume and a small opening. This makes it easier to construct the cover and to make the pit airtight, and results in a good balance between internal and external pressures. It also takes up very little land and is simple in construction, saving both work and material. At present this model has been popularised throughout the regions in Sichuan Province where stone is available as building material (Figure 4-1).

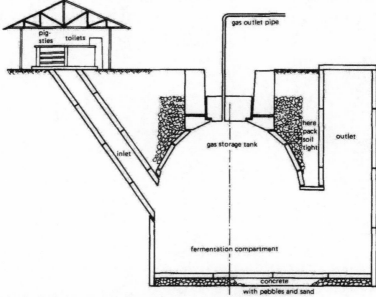

Fig. 4-1. Design of a circular pit made from stone slabs.

Design of the pit minimising the surface area
A domed biogas pit has a base diameter approximately the same as its height, and the shape of the pit is roughly spherical, which saves work and material. In order to increase the weight over the cover, maintain temperatures in winter, and guarantee the safety of the pit, the highest

*Properties of materials and cements used are tabulated in Appendix I.

point of the pit's dome should be 70-100 cm below ground level, so as to allow earth to be packed on top. To facilitate construction, a rough draft should be drawn beforehand in order to note certain points.

The diameter of the base of the pit is roughly the same as the height and this should be worked out from the required capacity of the pit. For example, a 10 cu.m capacity should have a 2.5 m base diameter. Scaling down, taking some fraction of the required base diameter, draw a square and bisect the square vertically, as shown in Fig. 4-2. With the Centre O of the square and taking OH as radius, draw arcs EH and FH. According to the planned length of the slabs of stone to be used for the dome, scale down and mark their lengths on the arcs EH and FH, as shown in the figure.

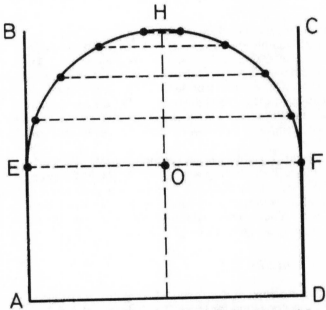

Fig. 4-2. Rough draft for a circular pit built of stone slabs.

Draw horizontal lines through these points, which will denote rings around the dome of the pit. The distances between these horizontal lines are the heights between the rings. The straight lines joining adjacent points give the length of the stones to be used, and the perpendiculars drawn from the points to the nearest vertical side of the square ABCD give the degree of incline of the rings. The horizontal lines joining the points give the base and top diameter of the consecutive rings; and at the top the distance between the two points on either side of H gives the diameter of the removable cover.

Calculation of the base width and top width of the stone slabs:

i. If all the stone slabs have the same width, then that width is the base width of the slab; and the sum of the number of slabs divided by the circumference of the upper circle will give their top width.

ii. If the stone slabs are of various widths, the base width of each slab is taken as the slab width; and the top width is calculated by multiplying the width of the slab by the ratio of the circumference of the upper circle over the circumference of the lower, giving the amount by which the slab width is to be decreased; dividing by two will then give the amount to be taken off each side.

To calculate the incline of these slabs of stone:

Subtract the diameter of the upper circle from the diameter of the lower circle and divide by two to get the horizontal distance by which the top is displaced from the bottom. For the design and method of calculation of a 10 cu.m pit see Table 4-1.

Materials needed in construction of the pit

For a circular stone slab pit with a capacity of 10 cu.m, 100-150 kg of cement are needed, 30-50 kg of lime and 300-500 kg of sand. The number of stone slabs should be calculated by the method discussed above. The thickness of the stone slabs depends on the size of the pit. For a pit of 10-30 cu.m the stone thickness should be 6-7 cm, for a pit of 100-200 cu.m the thickness of stone should be 9-10 cm. Taking a 10 cu.m pit as an example, the number of stone slabs of various sizes required is as follows:

Stone slabs: (length x width x height in centimetres) = (125 x 30 x 7) 25 pieces; (50 x 30 x 7) 25 pieces; (50 x 27.6 x 7) 25 pieces; (40 x 27.6 x 7) 20 pieces; (40 x 25.2 x 7) 15 pieces; long slabs (40 x 30 x 60) 6 pieces for the mouth; one circular stone for the cover (58 cm diameter x 20-30 cm thick). These measurements for the stone slabs can, of course, vary according to prevailing conditions.

Techniques in pit construction

Digging the pit: The diameter of the pit should equal the base diameter plus the width of the stone used for lining. The hole should be dug to the full diameter, straight down to a depth of 70-100 cm more than the depth of the pit itself, so that on completion the cover will lie 70-100 cm below ground level and earth can be packed down over it. This will increase the downward pressure and provide thermal insulation. If there is water at the foundation of the pit, refer to page 79 on the treatment of underground water and proceed as advised there. When the pit has been dug one should check the state of the soil and proceed accordingly. If the soil is firm and tightly packed stone slabs or irregularly shaped stones can be used and work can begin straight away. Lining them close up against the earth will save the work of extra digging, filling or packing in. If the earth is loose and soft, one may increase the diameter by 15-20 cm so as to pack on extra earth tightly to make a firmer wall.

the calculation of measurements for the construction of a pit of 10 cubic metres.

Height (cm)	Base measurements diam. (cm)	circum. (cm)	Top measurements diam. (cm)	circum. (cm)	Number of stone slabs required	Length of slabs (cm)	Width at base (cm)	Width at top (cm)	Len inw
123	250	750	250	750	25	125	30	30	
50	250	750	230	690	25	50	30	27.6	
43	230	690	184	552	25	50	27.6	22	
27	184	552	126	378	20	40	27.6	18.9	
19	126	378	58	174	15	40	25.2	11.6	

he bottom of the pit is 2.5 m; overall depth of the pit 265 cm; depth of foundation of pit 3.6 m; diamete

Pounding and paving the bottom of the pit: it is very important to pound and pave the bottom of the pit in order to achieve a solidity that will resist sinking, or water seepage in or out. If the bottom of the pit is hard — of sandstone or gravel — one may build the walls of the pit first and then the bottom. If the bottom is soft, then the foundations should be made first, then the walls, and lastly the bottom should be paved. This can be done with stones of 10-15 cm diameter or irregularly shaped stones packed closely together on the bottom; these should then be pounded down solidly. Keep pounding until there is no more sinking. After this, use a mixture of cement, sand and stones (ratio 1:3:8 by volume) to fill in the cracks and level the bottom. After the bottom has been sealed, it should be left for three days or more. When this paving and treating is complete, spread the bottom thinly with a mixture of cement and sand (ratio 1:2). One may also lay down slabs 8-10 cm thick to pave over the already pounded and cemented bottom, using a mortar of cement and sand (ratio 1:2) to cement them together. If the pit is built above sandstone or slate, the bottom can be simply lined with cement, then with slabs 8-10 cm thick. A centimetre or two should be left at the edges between the wall and the paving, and filled with cement.

Lining the walls of the pit: this is one of the main construction jobs and should be done as follows: prepare the stone slabs and put them in place around the edges of the pit, lining the walls, completing each ring to the right measurements before beginning the next ring above it. Each piece should fit into place firmly, so that it cannot move at all. Leave a space at the sides and bottom of each stone slab and fill it in with a mortar of lime and sand (ratio 1:2), spread evenly.

Where a pit is dug in tightly packed soil, the walls should be scraped flat and moulded into the right shape. They can then be directly lined with slabs or irregularly shaped stones resting firmly against the earth. Any cracks should be filled with a mixture of stones and a mortar of lime and sand. Line one ring at a time. To afix an inclining slab, put it in place and use a prop of wood or bamboo to hold it up; then go on to the next slab, and so on. When the last ring has been completed, cap it off with a slab or some small slabs together.

Where the pit is dug in loosely packed soil and pounding with extra earth has been necessary, the lining slabs should be filled in behind with damp earth as each ring is completed. This may be mixed with loose stones, and should also be pounded in tightly; do not pound in too much extra earth at once. It is best for two people to work on this together. When one layer has been pounded in, then start on another layer, and when the whole ring has been completed, start building the second ring. If the earth behind the slabs is not pounded in tightly enough, the walls of the pit will eventually move outward with the loose earth around the pit, causing cracks to form and impairing the pit's watertightness and airtightness. The tighter the earth is packed all round the walls, the better the pit will be able to sustain the internal pressure, and the stronger it will be. When the pit is completely lined, chisel a V-shaped gulley between the stones 1-2 cm wide and about 1-1.5 cm deep, to be filled later with mortar.

Lining the pit with irregularly shaped stones: After the bottom has been paved, line the walls with irregularly shaped stones one at a time. Each piece should be laid firmly against the wall of the pit. Once in place, it should be pounded in with a large mallet against the lining of the pit, and closely packed against adjacent stones. When one ring of stones is in place, then go over the whole ring again with the mallet and continue pounding until they are all firmly in place; none of them should move. Small cracks between the stones should be filled in with small stones. When this has been done all round, start on the next ring of stones. For mortar in this case, use a mixture of lime, sand and earth (ratio 1:2:1), or alternatively clay may be used. The lower portions of the wall may also be paved first without mortar, and the mortar added later. In building a pit with irregularly shaped stones, finish off the upper part of the dome with stone slabs, or use irregularly shaped stones throughout. In the latter case, for the final closing in at the top, the ring of stones should be slightly larger than the ring below, in order to push the centre of gravity of the stones back a little. To seal the opening, use small slabs or large irregularly shaped stones cemented together to make a cover. Whether irregularly shaped stones are used for the whole pit or only the bottom part (with slabs for the upper part of the dome), one must always be careful to pack the earth behind the stones very tightly, and earth should be filled in above the cover to a thickness of 70-100 cm, so as to increase the downwards pressure.

Lining the inlet and outlet compartments

The bottom of the inlet and outlet compartments usually lies about half way down the wall of the pit. The shape of the inlet and outlet may be either square or round, and the opening should be 40-50 cm across. For the upper part of these openings, slabs or irregularly shaped stones may be used to form an arch. The outlet should come up higher than the cover to the pit, and it should be able to accommodate 1-1.5 cu.m of liquid so as to prevent overflowing.

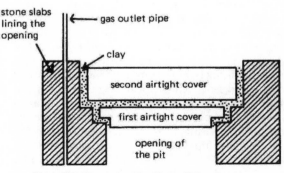

Fig. 4-3. The construction of the cover

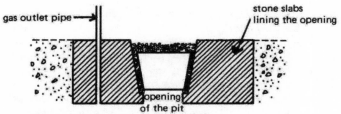

Fig. 4-4. (Alternative to Fig. 4-3).

Placing the removable cover

In the space reserved for this cover (50-60 cm in diameter), the earth should be levelled all around and packed tight, and stone slabs or ordinary pieces of stone used to line this area. The opening itself should be lined with one row of small slabs. In this row steps should be chiselled out to accommodate the actual cover. Two steps should be cut to accommodate a two-stage cover.

Around the cover itself there should be some removable material to function as a gasket (see Figures 4-3 and 4-4). The top of this should be flush with the ground. This opening may also be lined or paved conically. A conical shape should be hewn from rock and used as a top, sealed in with clay (see Figure 4-4). It is best not to have the gas outlet pipe going through the removable cover, but rather running beside it. When a complete overhaul of the pit is necessary, removal of the cover will allow sufficient light to enter, and after air has begun to circulate between the inlet, outlet and cover opening, any gas will be quickly dispelled. When scum is forming on the fermenting liquid, one may also open the cover in order to break through the hard scum and facilitate the production of gas.

Cleaning the inside of the pit

When all the above has been done, the inside of the pit should be thoroughly scrubbed with clean, fresh water, especially the cracks in the spaces between the stones; all sand should be washed off. After this, the whole pit should be lined once again with a layer of pure cement from the top downwards. A mixture of cement and sand (ratio 1:2) should be spread over this to fill in any cracks. The cement should be applied plentifully and made smooth. After filling cracks in the gas storage compartment and cover, there should be one more lining of cement and sand mixture (ratio 1:2), made very smooth, (this is not necessary for the fermentation compartment). This should then be painted over with the same cement, with vertical strokes and horizontal strokes alternatively, between three to five times, so as to increase the sealing strength. Although it is not normally necessary to do this in the fermentation compartment, it may be useful where the quality of stone is poor. When all this has been completed, care should be taken in maintenance. In winter a suitable temperature should be maintained. All fermenting material should remain in the tank for more than seven days before making any tests for water or airtightness (see Chapter 5).

Circular pit made of soft triple concrete*

In regions where there is no stone, triple concrete is an ideal material for the construction of pits. But because this concrete takes a long time to set, it is inconvenient to use it where it is damp underground. In building pits, people in Sichuan have improved the mixture of triple concrete, increasing its strength by mixing in coal cinders and stove ashes, which has speeded up the hardening process. A circular pit made from soft triple concrete with a large volume and a small opening is easy to seal and suitable for the areas where the earth is firm, the underground water level is low, and there is no water seepage. This construction is also quite suitable for plateau regions (see Figures 4-5).

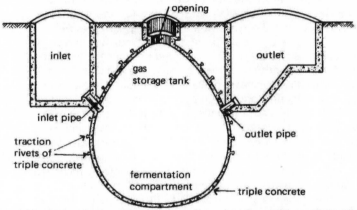

Fig. 4-5. Cross-sectional diagram of a circular construction triple concrete pit.

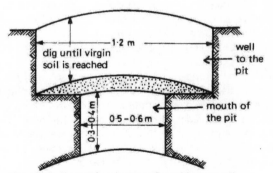

Fig. 4-6. A rough construction diagram of the opening to the pit.

*See note on page 28. The method described here for making triple concrete differs from the general formula in that plaster of Paris (gypsum) or cement is substituted for clay.

Construction plan

It is best to build this type of pit the way one builds sweet potato storage clamps, except that the opening should be a little smaller. This pit is suitable for all areas where sweet potato clamps are built. However, the capacity of these pits should not be too great, normally less than 15 cu.m.

Techniques in construction

One must first excavate the well to the pit (see Figure 4-6), 1.2 m in diameter, and deep enough to pass the cultivated soil and reach firm virgin soil. The circumference of the pit will be a ring with a width of about 30 cm, and the opening should be 50-60 cm in diameter. The mouth of the pit should be 30-40 cm deep.

Secondly, dig down from the mouth of the pit a further 1.5-2 m and then proceed digging according to the plan for the gas storage compartment. All around the walls of the pit, rings of holes 5-7 cm deep and 2-3 cm diameter should be dug, spaced out at intervals of 15 cm. When the pit is lined with triple concrete the concrete will enter the holes and form one entity with the pit wall lining and strengthen it. Before lining with triple concrete, make a very liquid triple concrete with 1 kg cement, 5 kg lime and 20 kg powdered coal cinders, and fill in all the cracks and crevices in the walls, pressing it down firmly with a steel spatula. Then apply a layer of lime mortar to the walls and tap it lightly twice over so that it combines with the soil. After this, plaster the inside with a mixture of cement, lime and coal cinders (ratio 0.1:1:6 by weight), to a thickness of 2 cm. When this layer has dried off slightly, apply another layer 1 cm thick, and press in firmly all over with a steel spatula. When this layer has dried a little, apply yet another layer of triple concrete. When this third layer has been pressed in firmly with the steel spatula, scratch lines all over the surface. These are drawn at about a 30° angle from the verticle, to the left and to the right of the vertical, so that the result is a criss-crossing of rhombuses on the walls. After this use a cement/sand mixture (ratio 1:1 or 1:2 by weight) to coat the inside twice, each coat 0.2-0.3 cm thick. After this apply two or three layers of pure cement mix.

Thirdly, when the gas storage compartment is completed, dig the lower openings of the inlet and outlet, and holes for the pipes that will link them to the gas storage compartment (see Figure 4-5), a clay linking pipe 1 m long and 15 cm in diameter should be inserted into these connecting holes and a fairly liquid triple concrete should be packed tightly around it to make it watertight. The cement lining of the other parts should be carried out as described above.

Quantity survey

For a 15 cu.m pit the materials needed are 100 kg cement, 200 kg lime, 600 kg crushed cinders and two lengths (1 m each) clay pipe 15 cm in diameter. The time taken to complete the pit is 30 working days.

40

Circular pit built with triple concrete bricks
In regions where the quality of soil is poor, or the underground water level is high and where the method described above is thus unsuitable, one can mould bricks from triple concrete. It should simply be slapped into brick moulds to produce bricks of a low specification number of grade. After that use the same methods as for the circular pit made of bricks.

How to make triple concrete bricks
The chief materials for these bricks are powdered or crushed cinders and lime and gypsum. The ratio of the plaster to lime to cinders should be 0.1:1:10 hy weight. These should be evenly mixed together when dry and then enough water should be added to make the stuff easy to mould by hand, so that the lumps disintegrate when dropped from a height of 1 m. This should then be forced into the brick moulds with great pressure, making a type of adobe triple concrete brick. These bricks should be left to dry for about 20 days through ventilation or wind, and should not be sun-baked or be rained upon. They will then be ready for use.
 For a 15 cu.m pit, 1500 of these bricks are needed, for which 25 kg plaster, 250 kg lime and 2500 kg cinders will suffice.

Points to watch out for in the construction of a pit
Follow the technique described in Chapter 4, page 44, on building a circular pit from bricks. Triple concrete bricks are not as strong as regular bricks; every square cm of triple concrete brick will only support a pressure of 15-20 kg (regular bricks will withstand a pressure of 100 kg per sq.cm). Therefore, with triple concrete bricks the thickness of the cover should be at the very least 24 cm, preferably 38 cm. The diameter of the pit should be less than 3 m.

Circular pit — a one-piece dry triple concrete cover
This type of construction was developed by combining the liquid triple concrete and the triple concrete brick methods. Its sealing qualities are good, the materials are easily available, and the capital cost is low. It is suitable for regions where the earth is fairly firm and the water level is low.

The shape and construction of the pit
There are two types: the round flat-domed, or round arch-domed. In order to seal and strengthen the pit, the triple concrete cover should become thicker with increasing distance from the opening, and should finally combine with the whole body of the pit. This will increase its resistance to internal pressure. A hole should be left in the centre of the cover piece for the removable lid and when fitted, this should be cemented in with clay.

Materials and labour necessary

The triple concrete used for this sort of construction is mixed from plaster, lime and cinders (weight ratio 0.1:1:10). A pit with a capacity of 20 cu.m should require under good conditions about 45 kg plaster, 450 kg lime, 4500 kg cinders, two clay pipes, one steel gas-outlet pipe, 35 man-days for the actual construction, and 20 man-days for transporting materials. Where the soil conditions are poor, the work required may have to be increased by 30-40%.

Digging the pit

Prior to starting work one should level the work site, choose a suitable central position and put in a stake from which measurements should be made and drawn on the ground.

Digging the pit cover

Make a mould by digging out ground in the shape of the cover you want to make, the size of which will depend on the required capacity of the pit. First, along the inside of the outer circumference line (drawn on the ground) dig a ring 40 cm deep; and then dig along the line defining the movable cover to a depth of 20 cm, then dig sloping outwards so as to leave a shape that is domed upwards in the centre. Then dig away the outer line completely so that the outside wall is slightly inclined outwards. Dig this ring quite deep, 60-80 cm, right down to where the walls of the pit will be, so that the cover piece will be linked to the walls of the pit in one single entity. The surface of this whole mould should be made smooth, the inclines should be made uniform in slope, the inner wall to the deep ring should be vertical, and the outer wall should have a slight slant. The central piece where the cover will be should be larger at the top than at the base (see Figure 4-7).

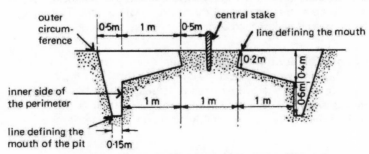

Fig. 4-7. A cross section of 3 m diameter, 2 m deep,
14.19 cu.m capacity, flat-domed one-piece cover pit.

Moulding the pit cover

It is very important that all the triple concrete should be poured in without interruption. Firstly, one must know the consistency and moisture content of the concrete. The mixing should be even and the concrete should be wet enough. After being poured in, it should be

pressed firmly. When mixing the dry materials according to the specifications, toss them around three or four times until the cinders, the
lime and plaster are evenly mixed. Then add the water until the cement
can be formed into lumps in the hand, which should disintegrate when
dropped from a height of 1 m. If it is too dry it will not mould properly,
and if too damp the quality will be impaired. It should be poured in as
soon as it is made, but first line the mould with old pieces of paper.
Start by filling the deep outer ring, then fill in the rest of the mould
and continue while pounding as required (the tighter the better). Keep
pounding right until some of the liquid comes to the surface; at this
point a good connection will have been made with the walls of the pit.
Two days after the cover mould is completely filled in, tap the surface
lightly with a flat object until the surface is made moist, then sprinkle
on a small amount of dry cement powder and spread this smoothly.
This should then be left for 20-30 days, and during this period it
should be shielded from the sun and kept dry; do not heat it artificially
or it will contract at too high a rate, causing unevenness and cracking.

Excavating inside the pit
After 20 days or so, when the cover has dried, dig down from the
mouth of the pit in stages, and after getting below the solid cover
piece, cement the sides. First, dig down 2 m deep from the top and
then dig outwards as far as the walls of the pit, which should be made
smooth and cylindrical. Then treat the cover for sealing and also the
upper part of the walls of the pit. When the upper portions have been
completed carry on digging downwards (as shown in Figure 4-8).

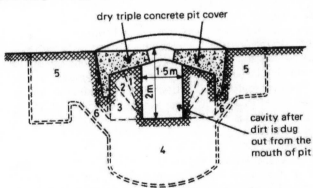

*Fig. 4-8. Excavation under the cover. 1-6 describes the order of exca-
vation of the pit: 1, 2 and 3 should be completed before going on to
4, then 5 and lastly to 6 (the connecting passages between the inlet and
outlet and the actual pit).*

Sealing the cover so that it will be air-tight
First, clear away any earth on the lower surface of the one-piece
concrete cover and the concrete walls of the pit, and remove the lining

paper and then apply a layer of pure cement. Next, apply a layer of cement and sand mixture (weight ratio of 1:2) which should be very liquid to give a thin coating. It should be applied with quick strokes all in the same direction, so that it will cement tightly. After two or three applications, apply another two or three coats of pure cement liquid of the consistency of thick soup. Then inspect carefully for any little holes or cracks.

Work on the bottom and walls of the pit
Where conditions are good and there is very little underground water, one can plaster the walls and pave the bottom with liquid triple concrete. First apply a coat of pure cement, then slap it all over with a flat object or a bunch of twigs. Do this three or four times to form a good bond between the cement and the earth outside. Then use a mixture of cement, lime and cinders (weight ratio of 0.5:1:3) to make a triple concrete with which to plaster the floor and the walls of the pit. This layer should be 1.5-2 cm thick, and again slapped in the aforementioned manner to form a good bond. Lastly, coat it with pure cement and sand mix (ratio 1:2 or 1:1) in a coat 0.5-0.8 cm thick.

Sealing the cover
First spread the inside of the mouth of the pit with a lining of clay 2-3 cm thick, then place the cover over this platform. Fix the gas outlet pipe and put a few large rocks on top of the removable cover to weigh it down. After this one may start testing with water or air to see whether the pit is up to standard.

Circular pit made of brick
In areas where there is little stone, pits may be built out of brick. This type of pit is suited to regions where the earth is firm and it can also be built in areas where the soil is sandy, but it is unsuitable for places where there is a shifting layer of sand or where the water level is high. The cost of such pits is greater, so it is best to make the maximum use of old and left-over bricks in order to lower the cost of construction.
    The construction method is as follows:

Digging a foundation
Dig according to the calculations and make the pit into a cylindrical shape. The walls of the pit should be smooth. If the pit is not made cylindrical and smooth, the pressure will be unevenly distributed, with a detrimental effect on the brickwork.

The order of construction
This should be based on the prevalent soil conditions. Where the earth is loose and apt to sink, one should first brick the cover piece, then dig downwards and line the walls with brick. Digging one stage after another in this way not only saves work and materials but also makes it easier to do careful work, and makes it possible to work in rain. When

digging down by stages, each stage should not exceed 2 m. Where the conditions are good and there is no danger of subsidence, and where the water level is low and there is no seepage of water on to the work surface, one can start by paving the bottom floor of the pit, and then go on to do the walls, lastly the cover.

Plate 4-1. Building the covering dome without support, using an arch base of ½ brick: the bricks are placed on their widest side, and the resulting dome is ½ brick thick.

Work on the cover

The work on the cover is the most important part of the work on a circular brick-lined pit. The arched dome may be built without support (see Plate 4-1). If this method is adopted, one should start with small pits and go on to make big ones gradually while acquiring more skill in bricking all the time. On no account should one start off by building a large pit with this technique, as it could easily lead to accidents. In bricking the dome, start with the bottom piece. For a pit with a diameter of 3 m, make a base one brick-length thick (hereinafter one "brick" = one brick's *length*); for the rest of the dome use a thickness of half a brick. For a pit of 4 m diameter, the base of the arch should be one and a half bricks thick; and for the rest of the dome one brick thick. For a pit with a cover diameter of 8 m, the base of the arch should be four bricks thick and the rest of the wall one brick thick. For a pit with a cover diameter of 20 m, the base of the arch should be eight bricks thick and the walls one and a half bricks. When bricking the base of the arch, the angle of inclination of the bricks should be

45

17°-19°, and the surface of the base bricks should be smooth (see Figures 4-9 and 4-10). In mixing mortar for the cover, the ratio of cement:lime:sand should be 0.2:1:3 by weight; the mixing of the mortar should be thorough, neither too watery nor too dry. In bricking the arch itself, the bricks should be made slightly dry so that they can still absorb water, which will facilitate bonding between bricks.

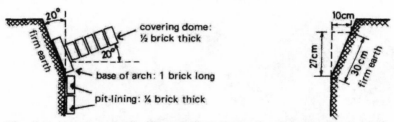

Fig. 4-9. Forming the inclined surface at the base of the arch for a pit cover of 3 m diameter.

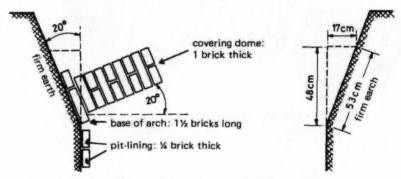

Fig. 4-10. Forming the inclined surface at the base of the arch for a pit cover of 4 m diameter.

In bricking, the mortar should be plentiful and the curvature in the structure and the angle of inclination should be maintained. The bricks should be closely packed. In successive rows of bricks the joins between bricks should not be above one another. In the slanting wall, a flat surface should be maintained. After every ring of bricks the mortar between that and the previous ring should be pressed firm by pushing with pieces of tile or flat stones. Then go on to the next row, continuing until you reach the mouth of the pit. From their experience in pit building, people summarise the main points thus: the mortar should be plentiful and spread evenly, neighbouring bricks should be placed close together, flat surfaces should be made properly, and the curvatures and angles of inclination should be precise. The quality of the pit cover depends on the laying of the bricks.

Fig. 4-11. Bricking a dome without support.

Fig. 4-12. Bricking the mouth of the pit without support.

The key to sealing the cover piece is the lining, or plastering. The plastering work should therefore be done very carefully. First of all the mortar between bricks should be pressed in, and the cracks filled up with more mortar. Before the earth is packed in over the pit cover, the mortar between the bricks lining the mouth of the pit and surrounding the cover should be partly scraped out and the resulting gulleys should be very carefully filled in with a cement:sand mix (ratio 1:2 by weight). Not a single gulley should be missed. When this is done, paint all over with pure cement (see Plate 4-2); over this, plaster on another layer of cement:sand mix (ratio 1:2 by weight), spread thinly so that the final thickness is 0.4-0.6 cm. The motions should be quick, spreading the cement on evenly. A very liquid cement mix results in a better bond with the brick. Finally, paint over the whole lining with a layer of pure cement, to a thickness of one or two mm, in a manner similar to the above mentioned. After this sealing is completed, wait eight to ten hours, and then apply another coat of pure cement the consistency of a thick soup. This should be repeated three or four times (see Fig. 4-11 and 4-12).

A spherical biogas pit made of brick, in construction; the surface is being coated with liquid mortar applied by brush. The inlet, visible on the right, is being dusted to remove loose particles. The outlet is visible in the background on the left.

Work on the bottom of the pit

In pits of capacity less than 2000 cu.m the lining to the wall of the pit should always be a quarter-brick thick, and the bottom of the pit should be flat. In bricklaying the mortar should be plentiful behind and between the bricks, and they should be laid so close that some mortar is actually forced out between them. All bricks should be closely packed, and the ends of bricks in each row should not match the places where bricks meet in previous rows.

Work on the inlet and outlet compartments

The bricklaying in these sections is similar to that on the bottom and walls of the pit. The opening to the inlet and outlet compartments should be shaped like an egg lying on its side with the small end closer to the fermentation compartment — this way it is more capable of withstanding pressures. There should be separate openings from the inlet and outlet compartments to the fermentation compartment. The opening from the inlet compartment should be larger than the opening to the outlet compartment, with diameters 20 cm and 10 cm respectively, either formed from clay or made with cement pipe. (See also Chapter 4, page 39 on inlet and outlet compartment brick-laying in the pits built from triple concrete and stones). In fitting the pipes from the inlet and outlet compartments, triple concrete should be packed very tightly all round the pipes so as to prevent any seepage of liquid. For an individual family pit the inlet compartment should be an ellipse with a major axis of 1.2 m and a minor axis of 0.8 m, and the outlet compartment should be an ellipse with a major axis of 1.5 m and a minor axis of 1 m.

Quantity survey of working materials (See Table 4-2).

Table 4-2.

Work, time and quantity estimate

Volume	Type of pit	pieces of brick	sacks of cement	kg. of lime	cu.m of sand	cu.m of cinders	man-work days
10	*	800	3	150	1	0.5	35
	+	1400	6	300	2	1	25
15	*	1000	4	200	1.5	0.8	65
	+	1900	8	400	3	1.5	85
20	*	1300	6	300	2	1	80
	+	2400	12	600	4	2	140
25	*	2000	8	400	3	1.5	180
	+	3800	16	800	6	3	250

* Large volume/small mouth
+ Cylindrical

NB: These bricks are regular bricks, and if half-sized bricks are used the numbers should be doubled.

Hemispherical or Pot-shaped pit

The hemispherical pit consists of a cylindrical fermentation compartment and a hemispherical dome, the whole shape resembling an upside-down wok.* The walls to this sort of pit should have a slight inclination, which may be adapted according to different soil conditions. The dome may be made of brick, stone or concrete, and supports are not necessarry. This simplifies construction and saves material. In the following we take for example a pit of 9 cu.m capacity, and the method is briefly summarised as follows:

Digging
Select a place close to a pigsty or toilet and dig a hole with a 3 m diameter. Dig it 1 m deep, then decrease the diameter by 0.5 m and carry on digging down for another 2 m, forming a stepped hole. The sides of the steps should incline inwards so that it tapers as it goes down, which facilitates dome construction. The lower section is the fermentation compartment. Its walls should be inclined according to the quality of the earth. The sides should be left rough, but any loose earth clinging to this surface should be brushed off, so that a good bond can be formed between this surface and the triple concrete that will be applied.

Lining the wall and laying the arch
Upon completion of the digging, line the sides of the hole with a triple concrete of lime, sand and clay (ratio 1:3:6). This mix should be pounded, made fine, mixed evenly, and then mixed with water. This triple concrete, after being plastered on, should be slapped all over, until it forms a rubbery layer. When this has been done, avoid sun-baking or wetting. When the triple concrete has set sufficiently, lay the bricks to form the dome according to the method described previously in Chapter 4, page xx on circular brick pits with an arch formed without support. The dome should be solid. Before building the dome the circular step on which it is built should be lined with triple concrete which should fuse with the triple concrete lining the fermentation compartment. To do this it must be pounded thoroughly. At the apex of the dome use a clay, porcelain or iron pipe as the gas outlet and cement it in firmly. When this hemispherical dome is completed, cover it with two layers of triple concrete, each 3 cm thick to seal it; in this mix the quantity of sand should be reduced slightly. Coat the outside surface with a mixture of lime and clay and slap it down. When the triple concrete has become hard enough, i.e. so that stepping on it will leave only a shallow footprint, cover it with loose soil layer by layer and tread on it, pressing each layer down firmly by foot. The total thickness should be no less than 50-60 cm. Final sealing with cement is even more effective; one may even plaster inside the dome with cement, repeating this several times every three hours until the surface is smooth and there are no holes.

*A Chinese frying pan, in the shape of a wide, shallow, concave dish, perfectly round. A pan of a very similar shape is used in East Africa.

The inlet and outlet compartments

The inlet compartment should open into the pit about half way up and should be 30 cm in length and width. The upper opening may be made a little bigger to facilitate the loading of material into it. The outlet compartment should open at the opposite side of the pit. The top should be elliptical and it should be built with three or four steps, the lowest three steps corresponding to the bottom, middle and top of the fermentation compartment. These steps should be rather wide, so that manure buckets can be placed on them when emptying or clearing out stuff from the pit, and so that a person can stand comfortably on them.

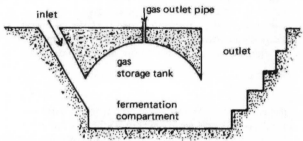

Fig. 4-13. Pot-shaped biogas pit.

Rectangular pit made of triple concrete

This is a type of pit adopted in places where sand, stone and lime are all easy to come by. The advantages of this type of pit are that the materials are readily available, it is easy to make at a low cost and the techniques required in the building are fairly simple and easy to learn. Here we describe two ways of doing it: one using dryish triple concrete which reaches that state by pounding, and a second using fairly liquid triple concrete.

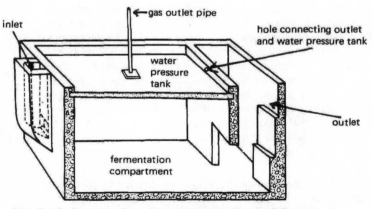

Fig. 4-14. Cross section of a rectangular pit of triple concrete.

Building a rectangular pit of dry triple concrete (see Figure 4-14).

Preparation of materials
To build such a pit with a capacity of 10 cu.m, 1250-1500 kg of lime, 200 kg cement, 2 cu.m sand and 6 cu.m of large and small stones.

Digging the Pit
When a suitable location has been selected, mark the outline of the pit and dig. The depth, which depends on the quality of soil and the water level, is normally about 2 m, and between 1.2-1.8 m wide. Where the soil is sandy or the earth is simply very loosely packed, a suitable incline should be left so that the pit tapers as it goes down — this is to prevent collapsing of the walls. Where the water level is rather high and there is serious seepage of water into the pit, the site should be moved or the length of the pit increased to allow for a compensatory decrease in depth.

Mixing triple concrete
The key is to control the water content and the sand content. Normally the suitable water content is 14% and sand content 63-75%. Clay has a low sand content and is more absorbent of water; sandy soil with its higher sand content is less absorbent of water. Therefore, in mixing lime with clay to make triple concrete, there should be an appropriate adjustment in the quantity of sand. If the sand content is too low, cracks will form all over upon drying, and this will decrease the strength. In mixing lime with sandy soil to make triple concrete, less sand should be added. In order to improve the strength of the triple concrete, a suitable amount of cinders may sometimes be added to improve the workability of the concrete. For a common triple concrete the ratio of lime:sand:sandy clay is by volume 1:3:3. There should be no twigs or bits of grass in the concrete, the lime should be thoroughly sifted, and the clay completely fine. Any hard pieces of lime should be removed, or they may absorb water and cause imperfections in the structure. A pit of 10 cu.m capacity will require roughly 8 cu.m of triple concrete. When the lime, sand, stones and clay have been mixed in proper proportions they should be stirred in their dry state and mixed thoroughly; then add water and mix thoroughly again. Add enough water to achieve a consistency where lumps can be formed which will break up when dropped from a height of one metre on to the ground. If the triple concrete has too high a water content, after it has been applied, it will contract as it dries and leave cracks; the result will be a failure to achieve watertightness and airtightness.

Lining the wall
The bottom must first be pounded very firm so that it will withstand the weight of the wall and the cover piece without sinking — sinking will cause cracks to form. Before lining the wall, draw a line on the

bottom to mark the desired thickness of the lining from the outer wall, and along this line put in some wooden stakes. Link these horizontally with some cross-pieces, put planks up against these stakes and the space left between the planks and the outer wall should be 15-20 cm. To form the new wall, the triple concrete should be put in between the planks and the outer wall in layers and pounded very firmly, so that it will set into a single piece (see Plate 4-3). It should also press firmly against the outer wall in order to be able to withstand pressure from within. When a layer of the wall has been completed, move the planks up. The horizontal cross pieces should be moved up as well and propped firmly against the stakes. Carry on building the wall, making sure that what has already been done is even, flat and not sloping.

Plate 4-3. Pounding the walls firmly in making a rectangular pit of triple concrete.

When building the wall, special attention should be paid to corners and places where different pieces of triple concrete join. At these places the moisture content on either side should be uniform and interlocking must be firm. Where two horizontal layers of concrete join, place a line of pebbles, approximately the size of a fist, along the centre of the lower layer, and the upper layer should press down upon this, to further strengthen the wall.

The wall should normally be built to a height of 30 cm or so below ground level to make it easier to replace the cover board or build toilets and pigsties. Where the water level is high and the planned effective capacity can not be achieved, the wall may be made somewhat higher, and may even protrude above ground level.

Building the separating wall and the outlet compartment
Between the fermentation compartment and the outlet there is a partition wall made of triple concrete and approximately 14 cm thick (see Figure 4-14). The two ends of this partition wall should be set into the walls of the two sides; only this way can it sustain the internal pressure from the pit. In the lower half of this partition wall and to one side there should be an opening to the outlet compartment. The opening should normally be half way up the wall lining of the pit. It should be 70-80 cm across to enable someone to go inside the pit to clean it out occasionally.

Inlet compartment
This should be a diagonal opening sloping into the central portion of the fermentation compartment. Its upper diameter should be no less than 50 cm, and on the average should be 50-60 cm. The lower opening of the inlet compartment and the opening into the outlet compartment should not face each other, as it would allow some of the material being let in to go straight out again. The mouth of the inlet compartment should be linked to toilets and pigsties (see Plate 6-1). The inlet compartment may be lined with triple concrete or stones, or it can be made of clay pipe. Whichever method you use, it should be built at the same time as the wall lining of the pit, so that a close bond can be formed between the inlet and the walls of the fermentation compartment for avoiding water seepage.

Cementing the bottom
After the walls have been built, a layer of triple concrete approximately 15 cm thich should be applied to the bottom and pounded firmly; this should be further covered with a coat of liquid triple concrete. If there is water seeping through the bottom, then a layer of pebbles should first be laid and then cement poured on top, and this should be covered with a layer of concrete (which is a mixture of cement, sand and stones with a volume ratio of 1:2:4). One must achieve a firmness in the bottom so that it will not sink; also it should not allow water through, and in fact must be watertight.

Building the cover
This will determine the success of the pit. In plains areas one can make this out of quadruple concrete (cement, lime, pebbles and sand) or it could be made of cemented pebbles or prefabricated boards of cement.

Covers made of quadruple concrete and pebbles are fairly cheap and effective. Where conditions permit, this type should be adopted. The methods of construction are as follows:

A simple arch-support: Take several wooden poles 12-15 cm in diameter and place them parallel on top of the fermentation compartment. Make them into an arch by placing bricks along two ends to support the poles. The height of the arch should be 20-30 cm. Over these parallel poles place some reed mats and tie them down with string. Then the arch is ready to be built. A similar simple arch support can be made from adobe bricks or pieces of wood, so long as the shapes resemble those described (see Figure 4-15).

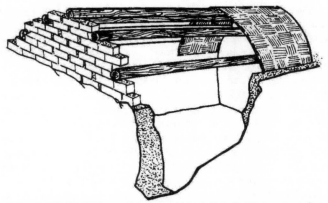

Fig. 4-15. Building a simple arch support for a rectangular biogas pit.

Arch made of quadruple concrete: Mix lime, cement, coarse sand and loose stones or pebbles 3 cm in diameter (ratio 0.5:2:2:4 by volume). First mix the dry lime, cement and sand evenly, mix in the stones or pebbles then and add the water slowly. When thoroughly mixed pour this concrete on top of the arch support to a thickness of 20 cm or so, then immediately rake it over with an iron rake and shovel, so that it lies firmly on the support. The stones or pebbles must be closely packed and the liquid mix spread evenly. When the concrete is slightly dry, cover this cemented arch over with damp reed mats. The mats should be kept damp by spraying with water for seven days.

Cover piece made of prefabricated concrete boards: Prefabricated concrete boards should be made from a mix of cement and sand and stones (ratio 1:2:4 by volume). Each piece will require around 25 kg of concrete. Before making the cement slabs you need to make wooden frames; they should be 60 cm wide, 50 cm thick, and be 30 cm greater than the width of the fermentation compartment. The frames should be made trapezoidal in cross-section — the measurements on top should be 1 cm larger than the measurements below. One may also dig a similar shape in level ground and use this cavity instead of the wooden frames.

Inside the wooden frames place some lengths of split bamboo as reinforcement. The separation between the pieces of bamboo should be 15-20 cm; they should be bound tightly with iron wire. Spread some very heavy brown wrapping paper with soap suds and put this on very flat ground. Lay the wooden frames on top of this and prop the bamboo reinforcement up with pebbles so that they lie in the centre of the wooden frame. Then pour in the well-mixed concrete to a thickness of 12-15 cm. This should be thoroughly prodded to make it solid and prevent it cracking. When these prefabricated slabs have been made, cover them with damp reed mats and leave for seven days; take the frame off only when the concrete is completely set. In placing the prefabricated cover over the fermentation compartment, first clean the tops of the wall lining the fermentation compartment and remove any grass roots. Coat it with a layer of cement:sand mixture (ratio 1:2 by volume). Put a couple of wooden sticks on top of the walls at either side of the fermentation compartment. Lower the prefabricated board down on top of these sticks, adjust until in position, and then remove the sticks. In this way one can avoid any severe vibrations or impact which could cause cracks in the concrete walls. Place the prefabricated slabs so that the wider, smooth side faces down into the fermentation compartment. Where the slabs meet the walls of the compartment, contact should be smooth and without cracks. The paper on top of the slabs should then be wiped clean leaving no clay or earth sticking to it. Pour a mixture of cement and sand (ratio 1:2) into the V-shaped cracks between slabs and fill them up completely. One may also make a simple arch support, according to the pre-described method; use cement, sand and stones for the concrete to form a one piece cover.

A cover made of round stones: (See Chapter 4, page 59 on the construction of rectangular pits with round stones). After one of the above four methods have been used to build the cover for the fermentation compartment, a further layer of liquid triple concrete should be used to cover the whole construction, to a thickness of 7-8 cm. This should be pounded and slapped repeatedly so as to make it very firm and increase the quality and weight of the cover. In the cover itself, two holes of diameter 3-5 cm must be left, one for the gas outlet pipe, and another as a safety valve. For the methods and uses of this latter, see the method of construction of a removable cover (see Chapter 4, page 76).

Building the water pressure tank

Around the cover of the fermentation compartment use triple concrete or pieces of stone to build a wall to form the water pressure tank. Normally the wall should be about 30 cm wide and 30-50 cm high and its base should rest partly on the pit cover and partly outside it, so as to increase the weight on the cover. At the base of the wall separating this water tank from the outlet compartment, a small hole (5-6 cm diameter) should be made to link the two. While building the water tank, also construct the outlet compartment or wall-lining out of triple concrete. The top of this wall should be level with the top of the water tank. In some places people have enlarged the outlet compartment, so that it serves the same purpose as the water pressure tank. Above the water pressure tank toilets or pigsties may be built.

Removing the arch support
Only after the pit cover has set and contracted can the arch support be removed from inside the pit. Once the supporting bricks at either end have been removed, the arch support will become loose and wobbly. Therefore, one must take special care while removing it to prevent accidents.

Coating the inside of the pit
Coating the inside of the pit is a crucial factor in ensuring watertightness and airtightness and great care should be taken with this work. Normally, there are two steps:

1. After the wall has been lined with triple concrete, apply a coat of coarse sand mixed with lime (ratio 1:1:5) 0.3 cm thick. When this has dried slightly, slap it or pound it firmly with a large mallet or a large flat surface. When the arch support has been removed, there may be some flaking away. If this happens, smooth over these areas and coat them over again. At the corners numerous coats should be applied, making the corners rounded.

2. Next, coat the inside with adhesive cement. Adhesive cement is made from cement, lime and sand (volume ratio 1:1:3). While cement and sand mixture alone, though firm, cracks when dry, adding some lime will increase the density and help make it watertight and airtight. Adhesive cement is normally applied in two coats, each 0.2 cm thick. Each coat should be smoothed over with a trowel three or four times. To further increase the wall's resistance to seepage, it is best to coat the areas lining the gas tank with a further three or four coats of a cement and water mix. This should be applied before the adhesive cement has completely dried — it will increase the airtightness.

Rectangular pit constructed from liquid triple concrete
This type of pit construction is economical in work and materials and is suited to regions where the soil is firm and unlikely to cave in or crack, that is, in areas with sandy soil. The method of construction is as follows:

Digging the pit
According to the dimensions of the plan, choose a suitable site and

draw the outline on the ground. The fermentation compartment should be elliptical in shape and normally about 2 m in depth. The outlet compartment should be dug downwards, starting 80-100 cm down. Below the wall separating the outlet compartment and the fermentation compartment, make an opening 1.2 m high and 70 cm wide, or simply use a prefabricated slab of concrete 10 cm thick as a separating wall. This slab should extend about halfway down the pit. At the other end of the fermentation compartment make an opening about 40 cm across to link the inlet compartment to the lower part of the fermentation compartment. The inlet itself should be funnel-shaped to permit easy passage of the material. If the inlet opening is too small, blockages may occur.

Preparing the walls
When the pit has been dug, clear away any stones, roots or bits of tiles.

Lining the bottom of the pit
First pound the bottom of the pit very firmly and then lay a floor of triple concrete made from lime, pebbles and clay (ratio 1:3:10), 15 cm thick. Pound it firmly.

Lining the walls of the pit
When the floor has been completed, start building the walls. If the lower half of these walls coincides with a stratum of gravel, then this region should be hollowed out some more and the difference filled in with a mixture of lime and clay (ratio 1:5) — the walls will then be even again. Then, using a mallet or stone, pound the walls firmly, and slap them smooth using wooden boards.

This should be done several times, so that the walls are quite smooth and without cracks. Next, all the walls of the pit should be scored with some sharp instrument, diagonally in both directions. The lines should be 0.5-1.0 cm in depth. Apply a coat of lime about 0.3 cm thick, and pound again to make sure the lime will seep through to the earth lining the pit and make it strong. After the wall has dried slightly, spread on a layer of lime and cinder mixture (ratio of 1:2); press this on firmly with a trowel over the whole wall, or use a mixture of lime, pebbles or coarse sand, and clay (ratio 1:2:2), which should be mixed evenly and applied forcefully on to the walls with a trowel. When this has dried somewhat, slap with boards until the surface is wet, so that the bond with the layer underneath is firm. Then apply a coat or two of cement, lime and sand (ratio 1:2:3). When this has been made smooth and allowed to dry a little, coat it over a few times with pure liquid cement.

Cementing the bottom
Use a mixture of cement, sand and round stones (weight ratio 1:3:7) about 6 cm thick. When the concrete has been poured in, leave it wet until it sets. Then coat all the pit walls and the bottom with a layer of adhesive cement, a mixture of cement, lime and sand (volume ratio

1:0.5:2). When this has been made smooth, coat it a few times with pure liquid cement.

Making the cover

In regions where such round stones are not available, prefabricated concrete slabs can be used to make the cover (see construction method for rectangular pits of triple concrete, Chapter 4, page 55). In laying the slabs, the wider surface must be placed (the smoother side) facing into the pit, and the places where the slabs meet the walls of the pit and the interstices between slabs must be filled with cement. When the cover is properly placed, the sharp corners where the wall and the cover meet should be made round by laying on cement. Also, where the gas outlet pipe extends into the pit, a protective shield or guard should be made out of cement.

The water pressure tank

Along the four sides of the cover, a low wall should be built about 30 cm in height and thickness, and at the same time the height of the inlet and outlet compartments must be raised to bring them level with the top of the low wall; at some point in the base of the wall separating the pressure tank from the outlet compartment, make a connecting hole. When all this has been completed, coat with a layer of cement mixed with cinders (ratio 1:2).

Rectangular pit made with round stones

Where round stones are abundant, this type of pit is very suitable. The building materials necessary may be collected over a period of time in people's spare time. In the Quanxi plains in West Sichuan where there are many river beds, this type of pit is frequently built. Because these round stones are not uniform in shape or size, the whole pit should be built without interruption; and while it is being built, attention must be paid to quality. The chief points of construction are:

Quantity survey

According to the size of the pit, have plenty of building materials ready i.e. lime, sand (which should not contain small pebbles or bits of rubbish), cement, round stones (approximately 20 cm long). To build a 10 cu.m pit you will need roughly 1000 kg lime, 200 kg cement, 8 cu.m round stones and 3500 kg coarse and fine sand.

The foundations and digging of the pit

According to the planned capacity of the pit and the size of the round stones to be used, outline the area to be dug, allowing for an inclination in the outer walls (the top measurements should be about 20 cm greater than those at the bottom of the pit) to prevent the walls caving in. When the digging has been completed, round the four corners and pound the bottom so that it is all quite solid. (The bottom should slope slightly downwards towards the outlet compartment.) Now line the bottom with round stones the size of tea plates (about 15 cm across). These should be pounded into the earth and smaller pebbles should be strewn in the interstices. Press these down firmly, then cover with a layer of triple concrete 5 cm thick, a mixture of lime, sand and soil (ratio 1:2:3). This layer should then be made smooth.

Building the walls
Use a lime and sand mixture (ratio 1:3) as mortar, and brick with large round stones packed fairly closely together — the mortar should completely fill the cracks to form firm bonds. In the second ring the stones should sit above and between each two pebbles below, so that the three stones form the three corners of a triangle. When each ring has been completed, pound every single stone tightly against the wall. The areas between rows of stones should be entirely filled with mortar, which can be mixed with small pebbles and earth. The opening between the fermentation compartment and the outlet compartment may be covered by an arch made of round stones or by a prefabricated concrete beam; then continue bricking up with round stones above this.

Lining the inlet and outlet compartments
These should be made with round stones using exactly the same method as was used to line the walls.

Building the arched cover
To make an arched cover out of round stones, an arch support must be built first (see Chapter 4, page 55 on rectangular triple concrete pits). It is best to use large rounded stones, slightly flattened and egg-shaped in cross-section. They should be washed clean and then placed with the smaller end pointing inwards, packed tightly in layers starting from either inside and working towards the apex of the arch. The cracks between the stones should be wedged tight with smaller ones.

Two holes should be left in the top of the arch, one for the gas outlet pipe and another as a safety valve. The safety valve is normally sealed tight, and only opened when large quantities of material are first fed into the pit, so as to avoid unnecessarily large pressures building up which could damage the walls of the pit.

When the bricking over is complete use a mixture of cement, lime and sand (ratio 1:2:5) to fill in all the cracks and crevices between the pebbles. Large cavities should be filled with small pebbles or bits of stone; the entirety should be whipped with bamboo poles or thin lengths of iron, as vibrations will help the cement to fill up the cracks. When the cover has set fully, cover over with a layer of liquid triple concrete or plain earth, to increase the resistance to pressure and strength and help prevent cracks forming as a result of exposure to the sun. When the arch has been completed, it should be left to harden completely before the arch supports are removed.

Plastering inside
This should be done in two steps. First cover the pit completely with

triple concrete to make the surface smooth. This layer should be pounded firm. When dry, coat with three or four coats of lime and sand mixture (ratio 1:2). Each coat should be applied with a trowel, without interruption to a thickness of 1 cm, and pressed firmly so as to bond tightly with the layer below it. When this is done, all the stones should be completely covered. Then smooth out the last coat. Secondly, on top of this last coat apply one or two coats, each 0.5 cm thick, of cement and sand (ratio 1:2) over the lining of the gas compartment. This should also be pressed in and made smooth. Finally, cover with two or three coats of pure liquid cement.

Lining the bottom
When the walls are finished, a layer of lime and sand mixture (ratio 1:2) should be applied to the bottom, 5 cm thick. This should be made firm and smooth, and special attention should be paid to the corners where the walls meet with the floor to ensure watertightness.

Rectangular pit built from long pieces of hewn stone
Such pits are very solid and durable and the techniques of construction are simple and suitable for any areas where stone is readily available. The size of this type of pit is normally limited by the span of the long pieces of stone used for the cover.

Quantity survey
To build a pit of 10 cu.m, 8 cu.m rough, unworked stone 1 m x 30 cm x 3 cm is needed, 400 kg of sand, 100 kg cement, 150 kg lime, one cu.m of loose stones, and one cu.m of stone slabs which measure 1.5 m x 40 cm x 15 cm.

Digging the pit
The pit should be dug 50 cm larger than the actual planned dimensions. When the digging is completed, pound the bottom firmly.

Laying the wall
First start with the corner stones, and lay the stones round in rings, layer upon layer. The interface between each layer should first be covered completely with a mortar made from lime and sand, and the vertical interface between stones in the same layer with a mortar of cement and sand so as to make the bond stronger. Each layer of stone should be fairly level upon completion. When a ring has been completed, place props between opposite sides, and fill the gap between the stone wall and the outer wall with clay. This should not be too damp, and should be pounded in firmly. The mortar used should have the following composition: for the fermentation compartment (the lower half of the pit) use a mixture of lime and sand (volume ratio 1:3) or lime and clay (volume ratio 1:1.5). For the gas compartment use a mortar of cement and sand (ratio 1:3).

Paving the bottom

When the wall has been built to over half its planned height, the bottom can be paved with irregular or regular shaped stones. When the bottom has been completed, the interface between stones should be filled tightly with cement and sand mix. Where the earth is fairly good, use a layer of triple concrete 8-10 cm thick.

Fitting the separating stone slabs

The wall separating the pit from the inlet and outlet compartments is usually made from stone slabs, 10-20 cm thick. The two ends of the separating slabs should slide into the previously hewn slots on either side (see Figure 4-16). After the slabs are in place, the slots should be filled up with cement and sand mix. The top of the separating slabs should come level with the top of the wall, and the bottom should reach down to about half way down the pit.

Where the lip of the outlet compartment is made (which is about half of the width of the pit), it is not necessary to make any special fixtures: the wall of the pit may be used as the separating wall. This will save the effort of hewing slots in the stone.

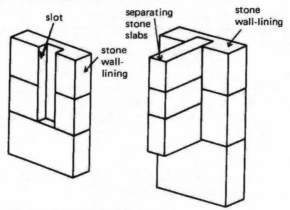

Fig. 4-16. Slots hewn in stone.

The inlet and outlet compartments

These should be made at the same time as the pit itself, and can be lined with either regularly shaped stones or cut stones. If we take the width of the pit and use this as the length of the top of the outlet compartment, then the top width of the outlet compartment should be 60 cm, and the walls surrounding the opening of the outlet compartment should rise 50 cm or so above the cover. If one makes the top width of the outlet compartment half as wide as the pit, then the length of the top of the outlet compartment should extend another 120 cm beyond the pit width, and a few descending steps should be made inside the outlet compartment. The inlet compartment should be a slanting trough or slide, with the lower opening at about half way down the pit.

A 60° inclination (from the horizontal) at the bottom of the trough will facilitate the inlet of materials.

Placing the covering stones

When the pit walls have been completed, and a step cut in the stone all round for the cover slabs, then take slabs 15 cm wide and fit them in. Where the long sides of neighbouring slabs meet, make the edges slant so that a V-shaped gulley is formed between slabs (see Figure 4-17).

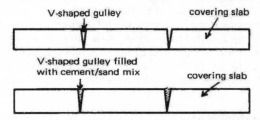

Fig. 4-17. The V-shaped gulleys between slabs.

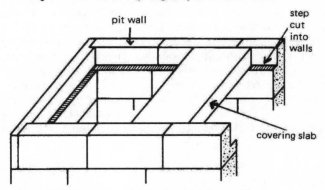

Fig. 4-18. The step cut into the top ring of stones round the pit.

When the slabs are laid into the steps, there should be a 1-2 cm gap at either end, which should be filled with cement and sand mixture. Where they rest on the steps in the walls, the surfaces should be covered with a cement/sand mix. In construction it is often best to leave one of these slabs unplaced, so as to let sufficient light in to see in order to complete the work, and this last slab should be put in place when all the work inside the pit is finished. The cracks at either end of the slabs and the V-shaped gulleys should be filled with small stones 0.3-0.5 cm long, and cement-sand mortar, to make the construction airtight.

The water pressure tank

The wall surrounding the water pressure tank may be built either from irregularly shaped stones or cut stones. This wall should stand right over where the covering slabs meet the pit wall, and should rest completely on a layer of mortar spread on top of the ends of the tank, and the wall. Should a water pressure tank not be required, then the upper portion of the outlet compartment (the portion of the outlet compartment which rises above the cover of the pit) should have a capacity of at least 1.5 cu.m.

The gas outlet pipe
This is normally made of iron. In a suitable place in the cover, make a hole somewhat larger than the diameter of the pipe and wash the hole clean, also file any rust off the pipe, stick the pipe into the hole so that the bottom end is level with the bottom side of the cover; on the upper side wedge the pipe tight with small stones. As a temporary measure, stuff a ball of paper in the bottom end of the pipe and fill in the crack round the end with fairly thick cement. When this has hardened slightly, start filling in the crack from above with cement mix, and build up a protective shield.

Lining the cracks between the stones
First, with a hard brush remove all earth and bits of rubbish from the cracks, then with some sharp object, scrape along the cracks to make a small gulley, and fill this with cement and sand mix (ratio 1:2) or cement, lime and sand mix (1:1:2); use a trowel to pack it in tightly. All corners within the pit should be rounded. Then put another line of cement about 1 cm thick in strips 10 cm wide along all areas where stones meet. Use a mixture of cement and sand (1:1) and make this smooth. When this has become fairly set, then coat the whole area, especially the gas compartment, with a pure cement mix. This should be left for seven days and then tests can be made by filling it with water.

Covering with earth
Where there is no water pressure tank above the cover, earth should be packed on to a thickness of 50 cm.

Matters to pay attention to when building this type of pit

1. One must have all the stone ready before starting to dig. Where the earth is of fairly good quality, the hole you dig should not exceed the planned capacity by much, because when packing earth back in it is hard to achieve the same firmness. As soon as the digging is completed, start building the wall immediately, to avoid collapsing or caving in the walls.

2. All holes, troughs and slots to be cut in the stone should be done first, before taking then into the pit. Do not try to do this work after the stone has been laid, as this may cause cracks in the mortar between stones as a result of vibrations.

3. If old stone is to be used, first chisel a fresh surface before putting the stone in place. New and old surfaces and interfaces must be washed clean, otherwise the mortar will not be able to form a good bond.

4. In the upper section (the gas compartment) of a rectangular pit made from stone slabs or long pieces of stone, one may introduce one or two props — horizontal cross pieces — to hold the walls apart against pressure from outside, and this can also be useful when there is a sudden movement of material into the outlet compartment, causing a drop of internal pressure which might damage the walls and the cover.

Building a pit in shale ground

It is very easy to build a pit in shale; it is also economical in work, material and money, and is used widely in Quxian in Sichuan Province. Normally for each cu.m only two to three work days are required, the cost is only 1-2 ¥uan (£0.30-0.60), and about 10 kg cement is required.

The surface layers of shale are usually fairly loose and the layers underneath are more solid. Therefore it is often best to build the upper half of the pit from, say, stone slabs, cut stone or irregularly shaped stones. In digging the hole, the slate is exposed to air and cracks easily and erodes. The small fragments that come off do not combine readily with cement, therefore one must seek to work very fast, digging, plastering, and filling in any cracks. The aim is to reduce the time taken for construction and cut down erosion. On all accounts avoid filling cracks up with loose stone, as all cracks must be filled very firmly. In order to ensure that they are well filled, normally apply three coats, first a small quantity of pure cement, then a coat of cement and sand mix (ratio 1:4) to give it strength, and then a third coat of 1:1.5 cement/sand mix to create an airtight seal. This is the key to the bonding of cement and slate. Methods of construction are as follows:

Digging the pit

The pit is normally oval or round in shape. When the location has been decided, have plenty of stone slabs ready or pieces of rock, and then according to the planned dimensions dig to form the pit. Any loose-lying layers of shale on the surface should first be removed. The faster the digging proceeds, the more you reduce the time of exposure and erosion.

Lining the wall

When the digging is completed, immediately place the stone slabs to form the wall of the pit and the partitions which create the inlet and outlet compartments. Stone surfaces should be refaced before cementing. In the lining, first use a lime and cement mix (ratio 1:3) to fill in the cracks, or 50 kg of unsoaked cinders, 15 kg of sand, and 7.5 kg of lime. Do not attempt to reface the stone once it has been cemented in, and do not try to fill up cracks with bits of brick, tile or stones.

Sealing cracks and plastering the walls and bottom of the pit

Any cracks in the walls, including those sections built of stone slabs, should first be chiselled into V-shaped gulleys and then filled with cement/sand mix (ratio 1:3). Then, over the entire area of the gas tank walls, spread two thin layers of cement, the same sort used to fill in the gulleys, and seal the cracks. This should be repeatedly smoothed and a strong bond should be formed. Total thickness should be about 0.3 cm; if thicker, flaking might occur. Lastly, give it a coat or two of pure cement. This should cover over any water marks and there should be no pock marks. If at this time the fermentation compartment walls and bottom are still dry, after the cracks have been filled in one can plaster this whole area with a kneaded mixture of lime and clay (ratio 1:4) which should not be more than 0.5 cm thick.

Fitting the cover board and water tank
See section on rectangular stone slab construction.

Pit carved in sheer rock
In large bodies of sandstone, sheer but not too hard, the surface layers erode easily. There are many advantages in building pits in sheer stone: the fact that the pit is built in one whole piece of rock means that it will be strong and able to withstand great pressure as compared to other constructions. The method is economical in its use of building materials, the skills required for construction are simple, and the cost is low. According to research done in Quxian, Zhongjiang, Jintang and Lezhi Counties in Sichuan Province, the average cost per cu.m for such a pit is around 1-2 ¥uan.

Fig. 4-19. A pit in sheer rock.

Selection of rock strata
The stratum chosen for construction of the pit should be fairly soft. Where the stone is harder the cost will be much higher. Before digging,

just remove the surface layer. Dig a small well in order to observe the lie of the grain of the rock. If the grain is horizontal, then the pit is not likely to collapse, whereas if the grain is vertical, slanting or irregular, then the pit walls may easily collapse or crack.

Shape of the pit
According to the geology and topography, a pit may be built to many shapes: spherical, cylindrical, elliptical, vase-shaped, sometimes boat-shaped, or even as a tunnel through a hillock. Rectangular pits are easy to build, but require more stone slabs for the cover, and building materials of other types.

Digging the pit
When a place has been chosen, first remove the surface layers of loose earth and stones.

1. For a circular pit that needs a cover but no partition walls, choose a suitable place to dig the main compartment. This should be dug down vertically but with a diameter of only 60 cm at the mouth, to accommodate a movable or fixed cover. When a depth of 60 cm or so has been reached, then dig outwards to expand the sphere of digging. The size of the main body should depend on the calculated capacity and the type of rock. Normally the final depth is about 3 m and the base diameter 3 m. Half way up the wall of the pit, dig a funnel-shaped opening on either side for the inlet and outlet compartments. The dimensions of the inlet will depend on the materials that have to be fed in; the outlet is normally 60-80 cm wide at the top opening and 150 cm long. A few steps should be carved on one side. Choose a location for the inlet and outlet compartments to suit the terrain and to allow for convenient inletting and outletting. Optimally they should be opposite each other (see Figure 4-20).

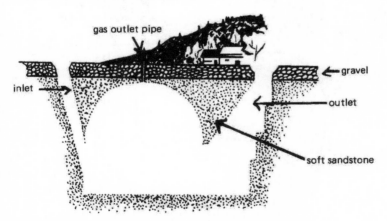

Fig. 4-20. Biogas pit carved out of soft sandstone.

2. A pit without cover board or partition may be in the shape of a boat or a pothole under hillock etc. Start digging from the outlet compartment: when you reach one half of the calculated depth for the pit, then spread out horizontally. Allow for the thickness of a partition wall, then carve out the gas compartment and the inlet compartment. The gas outlet pipe should be fitted near the top of the partition wall.

3. A tunnel type pit (see Figure 4-21) is built in sheer rock in the form of a tunnel and it has no cover. Partitions have to be built, but they can also be carved from the natural rock. First, dig the outlet compartment 1-1.5 m long and 80 cm wide and dig down to the calculated depth. Now dig the fermentation compartment and the gas storage compartment. This should be done by working forwards from the bottom of the outlet compartment, and then digging upwards. The shape and size of the pit should provide the required capacity and should fit the terrain or shape of the surrounding rock. Where the pit is being dug under level ground, the best shapes are cylindrical, round, or the shape of an urn. At the top of a hill or mountain, the digging should proceed parallel to the side of the mountain, to produce either a rectangle or a hemisphere. For a hemispherical tunnel pit the inlet compartment should be funnel-shaped and located more or less right across the pit from the outlet compartment. The mouth should measure 1 m by 50 cm. The lower opening should enter the pit at half the pit's height. The gas outlet pipe should be placed where the partition wall meets the ceiling of the pit.

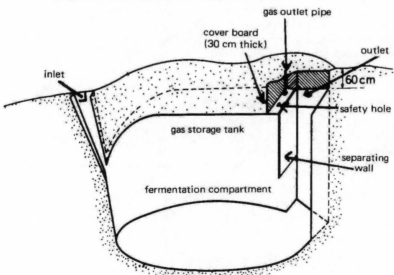

Fig. 4-21. Cave or tunnel-type biogas pit.

4. *Preparing the pit walls, filling in cracks, and plastering:* When the digging has been successfully completed, the walls should be smoothed out with flat chisels or some other instrument to remove any loose or crumbly layers and unevenness. Any cracks in the walls should first be chiselled into V-shaped gulleys and then water should be used to wash the entire wall and the cracks. Immediately fill in the cracks and start plastering. For the cracks, use a cement/sand mix (volume ratio 1:2) and pack it in solidly, pressing tight with a trowel. For cracks in the gas tank section, first apply a very thin coat of cement/sand mix (ratio of 1:3) 0.5 cm thick. Lastly, apply pure cement over the whole surface, rounding any corners.

Points to note during construction

i. Do not use explosives in excavating or tunnelling, because the vibrations may cause cracks to form, which may in turn result in water seepage and loss of airtightness.

ii. The top of the inlet and outlet compartments should exceed the ceiling or cover of the pit by about 50 cm.

iii. When the pit has been completed, if the walls of the pit or the cover show above the ground, they should be covered with earth in order to slow down erosion caused by exposure to air.

iv. After the cement has been left for several days, one may start putting water or material into the pit. A pit left empty and dry for a long time may slip, and all the walls may crack.

v. Throughout construction you must pay attention to safety and only proceed after careful investigation of the geology and the terrain so as to avoid accidents, especially through collapse or caving in.

Converting a manure pit into a biogas pit

Old manure pits of suitable capacity, whether round or rectangular in shape, may be converted into biogas pits. The method of conversion is either to add an inlet and outlet compartment to the original manure pit or to use some building material as a partition to separate off inlet and outlet compartments. All that remains is to make a cover and a gas outlet pipe. When converting old manure pits into biogas pits, the old walls and bottom do not need to be dug afresh. In fact, little needs to be done to them so long as they are strong. Thus, one saves not only materials and work but also ground area, so this is a recommended method. For the techniques required in the conversion, see preceding chapters of the book on various types of pit. Chief points to look out for are the following:

The inlet and outlet compartments: Their design may be of two types:

1. With the inlet and outlet compartments built outside the dimensions of the old manure pit. When a suitable place has been chosen for the inlet compartment, start digging. Half way down the wall of the old pit, dig a hole 40-50 cm in diameter to serve as the passage from the inlet compartment. Line the inlet compartment with brick, stone or tiles. At the other end of the old pit, dig the outlet compartment, as deep as the old pit, 80-90 cm wide and 1 m long. Then connect the lower half of the pit to the lower half of the outlet compartment with a fairly wide passage 1 m in height. The upper portions of the outlet compartment should be reinforced or lined; where there are arches they should be supported by stone or brick. The walls and the bottom of the newly dug outlet compartment may be made from brick, stones, large pebbles or triple concrete. This compartment should be slightly shallower than the old pit: if the old pit was not very deep and you find that the passage linking the old pit with the outlet compartment is not high enough, dig a trench along the bottom of the passageway so as to ensure that the height of the passage is no less than 1 m. This is to facilitate clearing out the pit.

2. The second case is where the old pit is rather large and one can build an inlet or outlet compartment at opposite corners of the pit out of stone or brick. The opening of the inlet compartment into the pit should be about half way up the pit wall; the outlet compartment at the other end is formed by building a wall or partition out of brick or stone, about 1 m from the end. This sub-divides the pit into two parts, a large volume from the fermentation compartment, and a smaller one for the outlet compartment. In the lower half of the partition wall, there should be an opening under an arch, about 1 m high.

The cover and the gas outlet pipe: If the old pit is circular, one may refer to the sections in this book relating to circular pits built of stone slabs or a brick arch. If the old pit is rectangular in shape, see the relevant sections on making covers out of triple concrete, round stones, or long stone slabs. If the old pit is wider than 2 m in the centre build two or three pillars for support; along the tops of the pillars one may cement on cross-beams of prefabricated concrete or stone slabs. The pit may then be treated as double; proceed in either half as before in the building of the cover. The gas outlet pipe should not be fixed until last. In converting old pits into biogas pits, there are two points to pay attention to:

1. Where the old and new material intersect or interface there must be a good connection.

2. Before plastering the walls, any old walls must be refaced by carving out a fresh face and then the new surface must be coarsened so that the plaster or coating cement will be able to bond the new and

old parts of the wall, so as to achieve watertightness and airtightness in the gas tank.

Building a Pit of Triple Concrete and Bittern*

Cement is an important building material, and where possible in the course of pit construction one should save it and use locally available materials. Cutting down the amount of cement required also reduces the cost. In various places where lime is easily available, people have invented a method using bittern on top of triple concrete to build the biogas pits, and their results have been quite good. They base their experience on the use of this combination for making threshing-floor surfaces, lining of chlocae, and surfacing of stoves and large water jugs.

The composition and properties of triple concrete

Triple concrete chiefly consists of lime, cinders, and sand or pebbles. The construction of the pit is similar to what has been described before, but there are two different mixing ratios; one for lime and sand and the other for lime and cinders.

1. The ratio of lime to sand should be calculated by volume, with more sand than lime. The fineness of the sand is also important. Normally, the base coats of the walls use cement made from a coarser grain sand, and the surface layers use a finer sand. The grain of the sand is what gives body to the material. Mortar made from coarse sand is tougher and more resistant to water when dried. Mortar for various building materials is mixed in volume ratios as follows: (i) under large pieces of stone the ratio of lime to sand should be 1:3; (ii) for use with large round stones in making an arch, the lime and sand should be mixed with little pebbles the size of small peas; (iii) for the cover to a rectangular pit made from triple concrete, lime and sand, and broken stone should be used (ratio 1:2:3); (iv) for plastering surfaces normally the ratio of lime to sand is 1:1.5 or 1:1.2 and there should be two coats.

2. The second kind of mortar is lime with cinders. Where cinders are amply available one may mix them, broken and sifted, into the lime and sand — the finer the cinders the better. The volume ratio of lime:cinders:sand should be 1:1:1. Lime and cinders can also be used as mortar for plastering the walls of the biogas pit. Cinders are a very good building material. They contain silicon dioxide and iron oxide, as shown in Table 4-3, which are classified as volcanic ash. Of these the silicon dioxide and aluminium oxide are free agents, existing as $Al_2O_3 . 2SiO_2$ or as amorphous free agents. Under the action of lime and water they can harden while wet. Powdered cinders and powdered lime with an admixture of 1-2% of gypsum is a cementing material which hardens in water.

*A black oily liquid residue from making salt, remaining in salt works after the crystallisation of the salt, consisting of magnesium chloride, magnesium sulphide, magnesium bromide, and sodium chloride; bitter and poisonous; commonly used in China for purposes such as those listed above.

Table 4-3. A chemical analysis of cinders.

Type	Chemical Composition (%)					
	SiO_2	Al_2O_3	Fe_2O_3	CaO	MgO	loss in combustion*
Cinders	47.72	28.21	12.24	5.22	0.75	1.66
Liquid cinders	45.95	32.53	17.97	4.54	0.47	1.94
Coal ash	44.85	33.78	11.70	5.27	0.55	3.26

*(as a % of the original weight of the coal).

The lime and uncombusted coal contained in the lime/cinder cement can undergo through heat treatment different chemical reactions similar to those undergone by ordinary cement. The reason why it solidifies is mainly due to the action of the silicon oxide and aluminium oxide, catalysed by lime (CaO) and water: various hydrates of silicon, aluminium and calcium are produced, causing solidification. For example, when Al_2O_3. $2SiO_2$ combines with lime $Ca(OH)_2$ dissolved in water and reacts, a complex of water, silicon, aluminium and calcium ($3CaO$. Al_2O_3. $2SiO_2$. $2H_2O$) is produced. In addition, the gypsum ($CaSO_4$), under the action of water, can produce hydrated sulphates of aluminium and calcium, which hastens hardening.

This lime/cinder cement is highly resistant to water because its calcium oxide content is rather low, unlike ordinary cement, which under the action of water will continuously yield calcium hydroxide. Therefore, this type of cement can be soaked in water when hardened and its chemical composition will remain stable. In fact, its hardness will often increase. This lime/cinder cement is also fairly acid resistant and alkali resistant.

The disadvantages of lime/cinder cement are as follows: (i) This type of cement when mixed with water is less exothermic than ordinary cement, in fact it only produces a quarter of the heat yielded by ordinary cement upon mixture, so it is not suited to working conditions where temperatures are below $0°C$. (ii) Because this type of cement takes a long time to harden, it is unsuited to regions where the water level is high and where there is seepage or accumulation of water during construction. (iii) It is rather porous and therefore should be pressed very firm during application and the surface should be made very smooth. (iv) In lime/cinder cement the uncombusted coal content should not exceed 5% of the cinders; if it is too high it will interfere with the qualify of the cement and decrease its hardness and resistance to water.

Coating with bittern

Rural people have much experience in applying bittern on to the surface of triple concrete, but the chemical reaction between the two has not yet been sufficiently studied. However, the basic chemical reaction and its consequences are described below.

Bittern is a by-product of the salt industry. Its chief contents are sodium chloride (NaCl), and magnesium chloride $(MgCa_2)$. The chief active ingredient of triple concrete is slaked lime $(Ca(OH)_2)$. The reaction between $MgCl_2$ and $Ca(OH)_2$ produces magnesium hydroxide $(Mg(OH)_2)$ and calcium chloride $(CaCl_2)$. The chemical reaction may be expressed as follows:

$$Ca(OH_2 + MgCl_2 \longrightarrow Mg(OH)_2 + CaCl_2$$

Here the magnesium hydroxide has no bonding effect and will not increase the hardness, but the calcium chloride will react with the $Ca(OH)_2$ and produce calcium oxide and calcium chloride salts. These salts are crystalline and increase the surface hardness of the triple concrete. The sodium chloride in bittern is, in fact, table salt, and a concentrated salt solution when applied to the surface of the concrete will extract the water from the concrete, causing it to seep to the surface. With applied pressure this can make the structure of the concrete much harder. Over all the places in the pit where lime/sand mix or lime/cinder cement have been used, bittern may be applied on the surface. Then use a rock or some other instrument to press the surface in strokes to smooth it out. This motion should be made repeatedly and always in the same direction.

In this way the capillaries or pockets of air in the cement can be forced out along with the formation of crystals in the tiny pores of the cement by the calcium oxide and calcium chloride salts. This will also increase the density and water resistance of the cement and thereby make the pit watertight and airtight. Because a smooth hard surface will prevent manure and other materials from seeping into the material, this will protect the walls. Normally for a pit of 10 cu.m capacity, 3-4 kg of bittern is needed.

The bittern should be applied when the cement or plaster has not completed dried i.e. when pressing hard with a finger still leaves a mark. The bittern should be diluted with half its volume of pure water to decrease its concentration. If the cement is hard and over dry or the concentration of the bittern is too high, the action of the salt will extract too much water and dehydrate it, thereby causing a bubbling on the surface and the whole layer of plaster may come off.

The smoothing action after the application of bittern may roughen up the surface and entire layers may even come away. When this happens, repair the damage by applying and smoothing more of the same mix but with some bittern added. Should cracks develop, they should be allowed to grow a tiny bit, and then repaired.

Building vase-shaped pits out of lime cinders and bittern
This type of pit has been built in clay-type soil of Meishan County of Sichuan Province, an old alluvial soil which is very dense and firmly packed, and which does not collapse easily. In this type of pit it is easy to achieve airtightness as both the requirement for material and the

costs are reduced and there is no need for stone slabs or large round stones. Here we describe the method of construction (see Figure 4-22).

Choosing the site
This choice of the site should depend on the structure of the soil, which should be firmly packed.

Quantity survey
For a pit of 10 cu.m, we need 450 kg of a limestone that will yield 70% or more lime, 600 kg fine powdered cinders and 100 bricks.

Outlining the pit and digging
First dig down 60 cm in the shape of a funnel with an upper diameter 1.3 m and a lower diameter 1 m — this will be used as the support for the cover to the fermentation compartment. Then continue digging down, and after another 30 cm start increasing the diameter and digging out. The ratios of the top, middle and bottom diameters of the fermentation compartment will depend on the density of the soil.

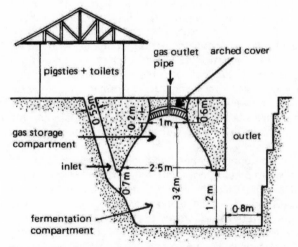

Fig. 4-22. Construction of a vase-shaped pit from lime, cinders and bittern.

For very dense, closely-packed soil the top diameter should be 1 m, the centre diameter 2.5 m and the bottom diameter 1.5 m, or in that ratio. For loose soil the ratio of top to bottom to middle diameter should be 1:2:1.5. Where the capacity is less than 10 cu.m the ratio should be 1:2.5:1.5 or 1:3:2, and for pits with a capacity of over 15 cu.m, 1:3:2.

The inlet compartment should be made in the shape of an inverted funnel with the bottom diameter larger than the upper. The diameter at the top should normally be about 60 cm and at the bottom 20-30 cm larger than this. The opening at the top of the inlet compartment should be 2 m away from the top of the wall of the fermentation compartment.

The outlet compartment is normally cylindrical in shape with the upper diameter about 50 cm larger than the lower diameter. Its capacity should be about one-third the capacity of the fermentation compartment. A distance of one metre should separate the top of the outlet compartment and the top of the fermentation compartment. The width of the passage linking the outlet compartment to the fermentation compartment should be 70-80 cm, and should be covered by an arch. When the digging is completed all surfaces of the pit should be made smooth to economise on material. In places where the soil is rather loose or damp, one may pound it and firm it with mallets or large stones and then proceed on to the next stage.

Building the walls
The walls are the key element in the construction of the pit. The cinders should be sifted very fine and the lime should be completely slaked (this will take one to two days). Any leaves and grass roots should be removed and the mixing should be done carefully with exact measurements. The moisture content should be just right. The mixture should be applied and pressed down firmly and the whole pit coated twice to increase the surface density of the walls. Both times, the weight ratio of lime to cinders should be 1:2 and mixed with water. The thickness of each application to the wall should be 1 cm. After each application pressure should be used to ensure that a good bond is formed between the lime and cinders with the wall, so that it does not flake. The surface need not be absolutely smooth. After about 15 hours apply another coat. This time make it smooth with a trowel so that it bonds firmly with the layer under it.

The bottom of the pit
First, mix limestone which is not completely slaked, or is left over, in equal quantities with the coarse cinders left over after sifting, and add some water. This should be piled up, covered and left for two or three days, so that the lime can become thoroughly slaked. Then this should be spread on the bottom and smoothed over, according to the method for the making of a floor of triple concrete, or slapped repeatedly, to bring moisture to the surface. Then one can apply the bittern two or three times.

Smoothing the walls
When all the plastering is finished, every few hours go over it with an iron trowel, to bring liquid to the surface and to smooth it. After this has been done two or three times with trowels, do it again two or three times with large stones. When the surface has partially hardened — when fingerprints leave no mark on the wall — then apply the bittern.

Building the arch cover

This covering arch is normally made of brick. The methods are basically the same as those used for the pits built out of brick which have been described. The requirements are that there should be plenty of mortar, and bricks should be laid tightly against each other. When the bricking over is complete use a mixture of lime and cinders (weight ratio 1:2) to fill all the cracks between the bricks, and press it in firmly. When this has dried, cover over with earth. Then plaster the inner surface of the arch with lime and cinder mix (weight ratio 1:2). Then apply bittern and smooth this out to make it airtight.

Putting in a removable cover

A removable cover was developed by the people of Sichuan Province to promote safety and facilitate maintenance of the pit. This removable cover fits into a hole made in the ceiling of the pit, round or square, and is normally about 40 cm across. One may lower a ladder into this to gain entry into the pit.

Three advantages of a removable cover

1. It will allow light into the pit and will facilitate maintenance and cleaning. When maintenance or cleaning is necessary, remove the detachable cover and there will be plenty of light to work by. Artificial lighting inside the pit is therefore unnecessary — this is what often causes gas inside a pit to catch fire, leading to burning accidents or explosions which can crack the pit. Once the detachable cover has been removed, the contents — manure and slurry — should be emptied out until the level drops below the mouths of the inlet and outlet compartment. In this way cross ventilation will carry off any remaining gases which will help avert any danger of suffocation to people going into the pit.

2. It facilitates mixing or stirring. When scum has formed on the surface of the fermenting liquid, one can remove the cover and use a bamboo pole or other instrument to break the scum or any hard surface to ensure normal gas production.

3. When large quantities of material are to be put in or taken out quickly, removing the cover prevents excessive pressure building up within, or a vacuum being created in the pit: either of these occurrences could cause cracks in the walls.

Method for attaching the removable cover

The removable cover for a circular pit is normally positioned in the centre of the ceiling and for a rectangular pit normally in one corner.

The diameter of a removable cover should be about 40 cm. It is best made in the shape of a stepped plug and placed into position with clay, or something very glutinous as a seal. One should then build a little mound in the shape of a ring enclosing the whole area above the installed removable cover and put water into it. If no bubbles rise it shows that the seal is airtight and effective. If there is an air leak, however, one must reseal the removable cover. When it has been sealed in place, the ringed area should be filled with water regularly to keep the clay damp so that it will not crack. The removable cover will be exposed to the same internal gas pressure from the pit as the ceiling itself, so that when there is 100 cm of water pressure (equivalent to 0.1 kg per square centimetre) for a circular removable cover 40 cm in diameter will be subject to a force of 125 kg; it should therefore be heavy enough to prevent the gas pressure causing cracks in the sealing of the pit while it is operating over the range of pressures calculated for it to withstand. When the cover is of insufficient weight, stones and bricks may be piled on top to increase the weight.

Fig. 4-23. Fitting a removable cover.

There is another removable piece in the cover, much smaller in diameter, called the safety valve. Its function is basically that of the removable cover. While the cover boards are being placed, drill a hole of 5-7 cm in one of them or in part of the cover. Then take a long, tapering stick 15-20 cm long and wrap it in long cloths, like a bandage, so that it can be used as a plug. Spread some rather glutinous clay or other substance around it and then insert it with some force into the safety valve. Here again build a ring of clay or some other substance and fill it with water to make sure that it remains airtight and to ensure that the clay does not dry and crack.

How to deal with underground water
Dealing with underground water is an important part of pit construction. If this water is not properly dealt with, it can cause a great amount of damage and much loss of work through repairs and going over work that was done before, and it will worsen the quality of the pit. In their

efforts during pit construction people have adopted a policy of 'avoid, guide, and block' and the results have been good.

Avoid underground water altogather. In choosing a site for a pit, make every effort to find a place where the water level is low or there is very little underground water. If possible the bottom of the biogas pit should be made higher than the highest water level underground. If the water table is rather high in the area and does not allow room for the original pit design, one may revise the design and make the pit shallower, or only half submerge it to avoid the problem of underground water at the very beginning, before even digging.

In places where there is a heavy rain season in summer and autumn, causing a high water level, one should avoid building and use the time to get the materials ready; take the opportunity of winter and spring to build and complete the pit. The underground water problem must be dealt with properly, and ample allowance must be made for the effect of underground water during summer and autumn.

The location should be carefully studied; one may investigate the water level of wells in the area. Avoiding the action of underground water will ensure the normal construction of the biogas pit. During digging, if great quantities of underground water appear, it is best to change the site and avoid unnecessary problems in construction.

'Guide and Block' — These two techniques should be combined when there is a direction in the flow of underground water. If water seeps into a pit that has already been dug, it should be guided to a low-lying area outside the pit. One must do more than merely block the water, because then it will accumulate in the vicinity of the pit and exert pressure against the walls and bottom of the pit. The greater the seepage into the pit and the higher the water level, the greater the pressures on the walls and the bottom of the pit. When weak points in the bottom and the walls can no longer sustain the pressure built up by the underground water, seepage will occur and necessitate repair. For these reasons it is necessary to adopt the method of both guiding and blocking.

Before blocking one must check the reason for seepage, locate precisely where it is occurring and then take remedial steps accordingly. In areas where the wall or bottom has not been cemented over and in areas where the seepage is occurring at a great rate, it is relatively easy to locate the source of the water, but where plastering or cementing has already been done or seepage is slow, some dry cement powder may be sprinkled on the suspect area: damp spots or lines will generally indicate the holes or cracks through which seepage is occurring. If one discovers a whole area of wetness, then this method is no longer useful. Over this area spread a fairly thin layer of cement evenly and then sprinkle on some dry cement powder and proceed as before, the wet spots and lines indicating the holes and cracks. One may also use lime mix for this method. Alternatively one may dry the whole area with a cloth and then observe where there are any wet spots or lines.

Dealing with underground water at the wall of the pit

Where any seepage has been discovered, although this may be infrequent in the dry seasons of winter or spring, one may suspect that there will be a great deal during summer and autumn. Before starting work on the walls, dig a circular horizontal ditch around the outside of the wall 8-10 cm wide, 5-10 cm deep and make a straight ditch leading away to a low-lying area from some point in the circular ditch.

Fill the ditch up with broken tiles, potsherds, and stones and cover over with rocks or tiles to avoid it silting up with earth, and then build the wall. In pits where cut stones or slabs are used, while building, make a trail running along the wall, or even all round the walls, out of small stones and potsherds. This will form a water resistant layer which will allow the water to drain down to the circular ditch at the bottom of the pit and so drain away. If there is still underground water seeping into the pit upon completion of the wall, one can open a vertical crack along the wall, and at the point in the crack where there is the most water, open a hole and use a little bamboo pipe to channel the water into a bucket. At the same time fill the crack with cement/sand mix (ratio 1:2) or use gluey cement (for preparation see page 8) to fill up the crack. When the cement has hardened, then according to the size and depth of the hole, shape a wooden plug of slightly larger dimensions, remove the little bamboo plug and immediately drive this plug in. It should go 3-5 cm into the wall. Then fill the hole from the other side with the gluey cement mix and smooth with trowel, pressing it in firmly.

The treatment of water at the bottom of the pit

There are different methods according to the seepage rate:

1. For a rather low rate (a daily seepage of 1-1.5 tonnes, or 40-60 buckets): where the seepage is low but is also coming through the walls, one should dig a small ditch in the shape of a cross in the bottom before paving. The ends of the cross should line with the circular ditch already mentioned. Of course, where there is no seepage through the walls, this does not apply, as there is no circular ditch. In this case one arm of the cross should lead off to a low-lying area and the ditches should be filled with stones and potsherds as described before. At the centre of the cross, dig a pit about 30 cm across and 40-50 cm deep and line it with stones, slabs etc. Allow the water to flow into this pit and ladle it out as fast as it comes in, thus bringing down the water level below the work surface. Then pave the bottom. When the walls have been lined and plastered and the bottom has been finished, barring the central area of the pit, take a piece of stone about 20 cm thick of the same dimensions as the pit and cement in place very quickly with mortar (i.e. cement plus a congealant, mixed in a ratio 1:0.5 or 1:0.8). In the centre of the stone slab drill a hole about 2 cm in diameter. Sink the stone into the bottom of the pit and as soon as it is set, pump or suck the water out of the hole. After allowing 10 minutes for the mortar to set, fill in the hole with the aforementioned cement

and also drive in a wooden plug about 20 cm long, slightly bigger than the hole. On top of this, lay a stone somewhat smaller than the central area of the pit to bring it level with the pit floor. The same glutinous cement should be used for mortar. On top of the stone put a layer of this same cement, or a cement/sand mix (ratio 1:2), in two or three coats.

2. When there is greater seepage from below, the method of disposing of the water is as above, but instead of digging a pit at the middle of the cross, situate it at the end of one of the ditches digging a 2 m long extension channelling the water into a lower place, and then dig a hole there about 1 m deeper than the bottom of the pit, so that the water may flow into it. The water should be continually ladled or pumped out of this water drainage pit so as to maintain the bottom of the pit as a dry work surface. Do not fill this hole in until the biogas pit has been tested for watertightness and airtightness and is fully completed. When finally filling the water drainage pit, use stones not earth, so that the water will be given a way out. In areas rich in underground water supplies this pit may be made into a well, which would be useful in the event of drought. When the cement of the water resistant layer has completely set, it should be well cured for a period with water which will increase its hardness. This will eventually make it more impenetrable to water and will prevent this layer from cracking. Curing this cement should be started neither too soon nor too late. When its surface is a light grey colour, use a watering-can spray to water its surface slightly. In the summer sun, cover with straw or grass and keep wet. After seven days of such curing, water may be allowed into the pit to help the cement to set further. Where the cover has not been topped with earth, continue a supply of water right up to the time that the cement sets hard.

Preparation and use of congealant, and blocking cement

Take a quantity of water, bring it to the boil and pour in copper sulphate and potassium perchlorate. Stir continuously. When all is dissolved, cool to 30-40°C, then pour it into the sodium silicate and mix; leave for half an hour. It is then ready for use. Prepared congealant should be stored in a cool place, in the shade.

Ratios by weight of the constituents of congealant:

Name of Material	Ratio	Colour
Copper Sulphate	1	Blue
Potassium Perchlorate	1	Orange/Red
Sodium Silicate	400	Colourless
Water	60	Colourless

Cement mixed with congealant sets very rapidly. It is preferable to test the setting time with a small amount before using. The fluidity of the mix should be adjusted to suit the worker and only then should you make up the quantity of cement plus congealant required. Ready-made congealant liquid (i.e. not mixed with cement) may be mixed with water. It is best to add sufficient water to achieve a congealing time of one to two minutes.

2. Glutinous cement mix is widely used for blocking purposes. It is made up of cement and congealant in the ratio 1:0.5-0.6 (it may also be 1:0.8-0.9). The hole or crack to be filled should first be chiselled. Cracks should be chiselled into V-shaped gulleys and holes should have their edges champfered; the sides of the holes and cracks should be roughened, washed clean, and the prepared blocking mortar should be shaped speedily to roughly the shape of the hole/crack to be filled. As it hardens, press in hard with the thumb to ensure a good bond. When this has been done sprinkle on some dry cement powder to test for seepage. If there is no seepage, then coat with a thin layer of pure cement and also a layer of cement/sand mix and on top of this do the coating/plastering. If the water pressure under the bottom of the pit is high, then as soon as the blockage is effected the pit should be filled with water to increase the pressure from inside to counteract the underground water and avoid damage of the repaired section by outside water.

Other methods of blocking
In their attempt to build these pits people have produced many effective methods of blocking.

1. One is blockage by cement, bittern and sand mix. Use one portion of cement, add to it one portion of fine sand, mix thoroughly and sieve. When ready for use mix with bittern, the weight of which should be roughly 40% that of the cement and the sand. Mix quickly and fill into the hole or crack. Press and smooth with a trowel. Then the congealing process may be hastened by the use of ash bags (these are described later); after this, plaster and cure until set.

2. Another method is blocking by means of cement, plaster of Paris and sand mix. First, heat the cement to 50-60°C, and add it to a mixture of an equal amount of sand plus half that much plaster of Paris. Mix this thoroughly, then add half the total weight in water. When this is stirred and mixed thoroughly it is ready to use. Again, ash bags may be used to hasten the congealing. Plaster over this.

3. The third method is using clay to effect the block. In the case of cracks and holes where the seepage is low, highly glutinous clay and lime may be moulded into strips or lumps and pressed into the holes and cracks. When clay is used, it should only half-fill the crack or hole when pressed in firmly. Fill in immediately with

cement/sand mix. This should be smoothed over with a trowel and, again, ash bags should be used. Once set, it should be coated.

4. The fourth method is to use fibrous paper to block the hole. First, use very coarse paper made from rather thick fibres. Roll this up tightly so that there is no hole in the middle, sufficiently large to fill up the crack or hole. Its length should be such that the roll fills half the depth of the hole or crack. Then use a thin wedge of wood to drive the paper roll into the hole or crack. Drive it in hard and then fill it in with cement/sand mix, pressed firm and trowelled smooth, then use the ash bag again and plaster over.

5. The fifth method is to use water pressure. If the water pressure under the bottom of the pit is high once the block has been effected, then immediately cover the area with a few pieces of old newspaper or a plastic sheet, pile clay on top of this and place a large rock over the whole. Then immediately let in water to fill the pit up to a level higher than the underground water level to increase the outward pressure from inside the pit. This will help the block resist the outside pressure and will enhance your chances of success.

6. The sixth method is blockage by means of absorbent ash bags, a popular and frequently used method. When the crack or hole has been filled with blocking cement it normally takes rather a long time to set. An ash bag placed on top of the cement will absorb a certain amount of the liquid in the mix and will hasten the process of congealing in the blocking material. Ash bags are easy to make. Use an old sock or cloth bag, or a bag made of straw and fill with ash, e.g. wood ash. Then sew up the bag and it is ready for use. The size of the ash bag will depend on the size of the area affected by seepage. If the cement has still not set after an ash bag has soaked up water, replace it with another such ash bag. The ash contained in the bags that have soaked up the water may be used in the pH control of the fermentation liquid in the pit later. Note: In place of ash, lime may be used.

5. Maintenance and Quality Appraisal of Biogas Pits

Having completed the construction, the quality must be strictly appraised. Only when tests show the whole pit to be watertight and the gas tank airtight can the pit be filled. Should the tests fail, the leak or leaks must be carefully located and repaired. When the pit is in use there should be frequent checks and maintenance. Action should be taken as soon as possible when problems develop. In this way the proper use and production of biogas can be maintained.

Common methods of quality appraisal
The quality appraisal should be done conscientiously and responsibly. Pressure tests must reach design pressures: there must be no slacking of standards and muddling through, or the result will be a waste of human labour and resources. A few common methods for quality appraisal are described below.

By observation inside the pit
When the pit has been completed, go inside to check the walls and the bottom for cracks or holes. With your fingers or a little wooden stick, tap various points in the pit; a hollow sound indicates that the plastered layer has come away from the wall. In these places the plastered layer should be removed and replaced by a new, fresh layer.

By filling with water
Fill the pit up to the cover and allow time for the walls to absorb water and become saturated. When the water level has settled, mark the level. If the level has not dropped after a day, this indicates that the pit is watertight — otherwise it is not. If it is watertight, you may proceed to check the gas tank for airtightness. The following applies to pits where there is a water pressure tank above the gas tank. Attach a tap or valve to the gas outlet pipe (where the valve is attached to the pipe should first be checked for airtightness); then open the valve or tap and block the linking hole that links the water tank to the outlet compartment. Now bring up some water from the pit — it can be emptied into the water pressure tank. Extract water so that the level in the pit falls 40-50 cm below the cover board. Then shut the valve and start adding water again into the pit. When the water in the outlet compartment has risen above the level of the cover board then stop adding it. Allow it to settle and mark the water level. Now carefully check all interfaces, corners and junctures, for example where the inlet compartment joins the gas tank,

for bubbles rising to the surface of the water. The period of observation should be fairly long. If at any point there is a stream of bubbles it shows that air is leaking and the place should be marked for future repairs.

Checking with a manometer

1. Making a manometer: A manometer is a tool used to measure the pressure inside the biogas pit; it is simple in construction and easy to make (see Figure 5-1). Take two glass tubes 1-1.5 m in length with an internal diameter of 1 cm. Fix them to a board or a wall inside a room and join the bottom ends with a length of rubber hose or plastic tube; by the side of each tube make graduations in centimetre units. At the top of tube B, fix a round safety ball, or a bottle without a bottom of more than 200 ml capacity. Fill with coloured water (to faciliate observation) up to the level of the zero mark. Take a 'Y' tube and join to tube A, and observe the change in the water column in the U-shaped tube (which consists of tube A and tube B and the hose). From this we can tell the pressure inside the pit. For every 10 cm difference in the water levels on either side there is a corresponding change in pressure of a hundredth of an atmosphere. For example, if the level in tube A were to drop by 20 cm and the water level in tube B were to rise by 20 cm, then the difference in the level in the water column will be 40 cm — this will be called an internal pit pressure of 40 cm water.

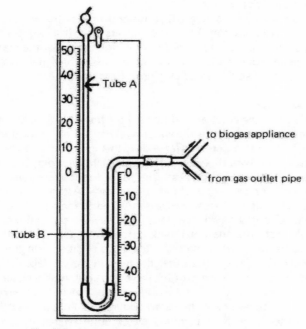

Fig. 5-1. A manometer.

A manometer not only checks the pressure, but also tells the magnitude and all pressure changes inside the pit. Therefore, one can control the pressure and ensure safety of the biogas pit. When the pressure within the pit is too high the water in the U-shaped tube will flow up into the safety bowl or bottle, and allow excess methane, or biogas to escape through the safety valve, thus automatically reducing the pressure within the pit. And when the pressure within the pit has been reduced to what is tolerable, the water will once again flow into the U-shaped tube and so maintain the gas pressure within the safe zone, avoiding damage to the pit from excess pressure within.

2. Using the manometer to check for watertightness and airtightness: Before attaching the manometer, first open the valve of the gas outlet pipe and fill the pit with water stopping when the inlet and outlet compartments are roughly half filled. Leave for three to five hours until the pit walls have become saturated with water. Then note the water level. After a day, see if there is any change. If there is any significant drop in water level, this will mean that the walls or bottom leak. When the water level stops dropping, make a mark on the wall. This will tell us that the leakage takes place between the initial level and the lower, final level. When leaks have been mended and you have made sure that the bottom and sides are watertight, then connect the manometer to the gas outlet pipe and start adding water, or else pump it up with air, using pumps or chemical spray devices. The purpose is to increase the air pressure inside. When a noticeable differential has been reached in the manometer, stop adding water or pumping air. Leave for 24 hours. Observe whether there is any drop in pressure. If the change in the height of the water column is nil or very small, i.e. 1-2 cm, the pit is shown to be airtight. However, a great drop in the water column indicates that it is not airtight. In the pressure test, one does not normally build up pressures greater than 100 cm difference in water level (or whatever the designed difference should be), so as to avoid damaging the pit.

One may also test for airtightness by filling the pit full of water, fixing on the manometer and then extracting the water out of the pit, thus building up a negative pressure or vacuum.

Checking with smoke

Proceed as for the manometer check, only in filling up the pit stop the inlet of water before the water has quite filled the inlet and outlet openings into the pit. Take an earthenware basin and fill it with dry straw or wood shavings and light them. Then cover over with fresh green grass, sulphur powder, or some other smoke producing material. When thick smoke starts to rise, put the bowl through the outlet opening so that the smoke rises into the pit. Then quickly raise the water level to seal off the inlet and outlet openings. After that, reduce the rate of water inlet. Bring the water level up to produce the pressure for which the pit is to be tested. Since this method allows one to tell if there is an air leak and the rough position of the leak in the cover or tank, it is better than a mere check with a manometer.

Checking by plastering
This method is useful for pits that have still been shown to leak after water and air tests have been made; and will help locate those leaks. Plaster over the walls up to and including the level at which leakage no longer occurred. When this lime plaster is almost dry observe whether there are wet spots, patches or lines. Mark these places and repair them later. If you are using this method for pits that have already been used to produce gas — where material has been inletted — it is important to clear out all the fermentation material and residual gas before remedial work is started, to avoid danger of suffocation, poisoning or accidents.

Leakage tests for the gas outlet pipe and gas hose
Before using the hose it should be thoroughly checked with the valve attached. Coil the hose up and seal one end by tying it tightly with rope. Then immerse the coil in a basin of water. Using a pump, or by mouth, blow air in from the open end. Check for air bubbles where the hole joins the valve, or where lengths of hose are joined. If there are no bubbles it is airtight. One may also apply soapy water with a brush to the valve and hose connections to check for leaks which would be indicated by bubbles. Alternatively hold duck or goose down close to these connections; if there is any air leak, the down will be blown by the air. This will locate the leak, which should be marked and repaired.

Causes and common locations of water and gas leakages
1. The foundations or the walls of the pit have not been pounded firm. In addition to water leaks this may even cause the walls to slope and the bottom to sink.

2. Leakage between stones due to the stone not having been properly cleaned before use, or the cracks having not been completely filled with mortar and pressed tight.

3. If after completion the pit suffers from considerable vibration, or if there is an earthquake, mortar or plaster can come loose. Alternatively, it could be that after plastering the pit was not properly guarded against the sun or rain, which caused cracks to form while the cement set.

4. The cement or sand used contained too many impurities.

5. Leaks round the gas outlet pipe, caused by rust which was not removed at the time of installation. Any particles of rust will prevent a good bond between the cement and the pipe. Or else it could be that when the gas outlet pipe was being fitted, the hole in the cover board was made too small or too smooth, or

that the hole was not washed out properly, or it suffered some vibration before the cement set; all these can cause gas leaks where the gas outlet pipe goes through the cover.

6. Leaks between the walls of the pit and its cover can occur if insufficient mortar was applied when the cover was fixed in place; or if too much force was used when the cover was being fixed and all the mortar was squeezed out; or if there was any vibration after the cover was successfully fastened on and plastered.

7. The cover may have been incorrectly plastered, allowing the gas to leak by capillary action; or the interfaces between prefabricated concrete slabs and stone slabs may not have been chiselled into a V-shape. These could cause gas leaks.

8. For pits made of triple concrete, the ratio of lime to sand, or water to lime, may not have been ideal, in which case severe contraction on drying would have allowed cracks to form. It could also be that the concrete was not mixed evenly; or that the lime was not thoroughly slaked, or even that there were admixtures of small pieces of unslaked limestone, which when built into the wall absorb water and disintegrate. Maybe not every layer of concrete applied was pounded tight, or the plastering may not have been of a high enough quality, so that no proper bond was formed between layers and for this reason plaster flaked off. Another possibility could be that the plastering with cement was not immediately followed by a period of protection and the pit suffered from sunshine, sunbaking, or from frost or rain. All these can cause cracks to form. It could also be that the earth which was filled in outside the wall was not pounded in sufficiently tightly so that the walls warped and caused air or water leaks.

9. It could be that during water tests or air tests when a new pit was filled with water by a pump, it was filled too quickly, and after the water level rose above the opening, from the outlet and inlet compartments, the gas outlet pipe in the cover was too small in diameter and air could not escape fast enough: this would cause a large pressure to build up inside the pit which in turn could cause cracks to form in the ceiling or walls. Alternatively, during long periods of use when vast quantities are inletted and outletted it could be that this was done without detaching the gas hose; over a short period of rapid intake or output of contents, positive or negative pressures can build up which may damage the pit.

10. If a newly-built pit is filled with materials before it has had time to cure, or if the site was not properly chosen, for example if tree

or bamboo roots have penetrated the walls of the pit, both events might lead to air or water leaks.

11. In areas where the underground water level is high: after out-letting material in summer or autumn — the wet season — if the pit has not been immediately filled with material or water, the underground water can damage the bottom and cause leaks.

12. Pits built out of stone material: if the earth has not been pounded firmly outside the walls of the pit during tests or during gas production, unequal external and internal pressures can make some of the stone building blocks move, leading to water or air leaks.

How to Repair the Biogas Pit

When an air or water leak has been located it must be carefully repaired. If large scale repairs are needed, all the contents of the pit should be removed and the pit cleaned. Take special care not to enter the pit before all the gas has been dispersed and there is plenty of fresh air inside, in order to avoid accidents from poisoning or burning. Methods of repair include the following:

1. Use cement and sand mix (ratio 1:1) to fill in the leaks and then cover the repaired area with three or four layers of pure cement. Cracks should be chiselled wide open, the edges roughened and filled in with the mix, trowelled smooth and carefully protected. For stone pits: before repairs, any stone surfaces and cracks should be chiselled clean. Any cracks in triple concrete pits should be opened up wide and deep, and then filled with cement, lime and cinder or sand mix (volume ratio 1:2:3) and this should be pressed tightly into the cracks and made smooth.

2. Any plaster flaking off in the gas tank should be cleaned off and new plaster applied.

3. Where an air leak has not been clearly located, the gas tank should be washed clean and a coat or two of pure cement, or cement/sand mix (ratio 1:1 or 1:2) should be applied.

4. When there is a leak at the point where the gas outlet pipe goes through the cover, the interface between the pipe and cover may be chiselled open around, and the pipe cemented in anew. Another alternative would be to enlarge the cement base, or guard, around the bottom at the lower end of the pipe.

5. Where there have been water leaks in the bottom of the pit which have been caused by underground water; see Chapter 4 page 77 on dealing with underground water. When subsidence of the pit

bottom has led to water leaks, whether the whole bottom has subsided or whether the cracks are at the angles where the bottom meets the walls, first enlarge the cracks round the walls by chiselling a gulley 1.5-2 cm wide and about 3 cm in depth. Do this all round the pit and then pound concrete to a thickness of 3-4 cm over the whole bottom and in the encircling gulley, so that it can solidify in one piece. When there are leaks in the bottom of a concrete pit, the bottom may be reinforced with a mixture of small pebbles and triple concrete, or with cement/ sand mix.

6. In pits built of stone: if there are cracks in the walls which leak water or air, they should be chiselled into V-shaped gulleys; then hammer a steel or iron plate into the cracks to make a tight fit between stones. Wash out the cracks and fill with cement/ sand mix. This should be carefully pressed in and smoothed; cure until properly set. In a circular pit built with stone slabs, if the earth is too loose behind the stone slabs, you cannot repair it by the steel or iron plate method. One should make 20 x 30 cm rectangular openings at the position of the crack, between neigh-bouring stones, and pound in earth behind these stones with stout wooden stakes. Pound in all directions, up, down, to the left and right — repeatedly. Fill the cavity formed with earth and stones and repeat the pounding and filling until quite tight. Then seal with a concrete or cement/sand/stone mix (ratio 1:2:4) and plaster over.

6. Scientific Management of a Biogas Pit

Upon the completion of a biogas pit, conditions have been created for holding fermenting materials and for producing and storing gas. However, the internal mechanism necessary for the production of biogas is a breaking down of organic materials by fermentation by the anaerobic and methanogenic micro-organisms in suitable living conditions.

Therefore, to maintain the normal fermentation process and fulfill the users' needs throughout the year, one must manage the pit scientifically. Practice has shown that there is a great difference between conscientious and careless management. When conscientiously managed, pits of small capacity will produce much gas and ensure supply. On the other hand, pits of large capacity, poorly-managed, may produce less gas than smaller pits, or may not produce gas at all, and will certainly not guarantee a normal supply.

The biogas pit has a function in the daily life of the commune member, but it is also of use in agriculture. The scientific management of the biogas pit must be geared to agricultural production i.e. the use of gas, the collection of fertilizer and the treatment of sewage must be properly integrated in order to provide fertilizer for collective production and gas for domestic needs. In the proper management of biogas pits, the following points should be noted:

Mixing material for fermentation
In rural areas many materials may be used for gas production — these are everywhere: human and animal excrement, stalks and foliage, grass, vegetable stalks, garbage, sludge, and industrial waste with organic chemical content are all good raw materials for gas production.

We know from the second chapter that one important factor for normal fermentation in the biogas pit is the carbon/nitrogen ratio in the material used. The ideal carbon/nitrogen ratio is around 20-25:1. Different materials have different ratios, and even the same material may have a different ratio under different conditions. Some common materials for fermentation have ratios as shown in Table 6-2.

Therefore, at time of inlet of material, not only must one have a definite quantity of fermentation material but one must also pay attention to the ratio of the various fermentation materials in the pit. There should be suitable carbon/nitrogen ratio, especially with materials of high fibre content, such as stalks and grass, and materials of high nitrogen content such as human manure. These must be used together

Table 6-1. Gas yield of some common fermentation materials.

Material	Amount of gas produced per tonne of dried material in cubic metres	Percentage content of methane
General stable manure from livestock	260-280	50-60
Pig manure	561	
Horse manure	200-300	
Rice husks	615	
Fresh grass	630	70
Flax stalks or hemp	359	59
Straw	342	59
Leaves from trees	210-294	58
Potato plant leaves and vine etc.	260-280	
Sunflower leaves and stalks	300	58
Sludge	640	50
Waste water from wine or spirit making factories	300-600	58

Table 6-2. Approximate values for the carbon/nitrogen ratios of some of the common materials used for biogas pits.

Material	Carbon as a percentage of total weight %	Nitrogen as a percentage of total weight %	Carbon/ nitrogen ratio %
Dry straw	46	0.53	87:1
Dry rice stalks	42	0.63	67:1
Maize stalks	40	0.75	53:1
Fallen leaves	41	1.00	41:1
Soya bean stalks	41	1.30	32:1
Wild grass: i.e. weeds etc. (in China often narrow, thin leaved)	14	0.54	27:1
Peanut vine stalks	11	0.59	19:1
Fresh sheep manure	16	0.55	29:1
Fresh cow/ox manure	7.3	0.29	25:1
Fresh horse manure	10	0.42	24:1
Fresh pig manure	7.3	0.60	13:1
Fresh human manure	2.5	0.85	2.9:1

with other materials. Practice shows that fermenting a single kind of material generally gives poor results. In some areas people have kept roughly to the following proportions: 10% human excreta (including the liquid fraction), 40% pig, cow and animal manure with stalks and grass, and 50% water. With this mixture the result is fairly satisfactory. Since there are various sources for fermentation material in rural areas, naturally one must not be rigid or pedantic. A suitable mixture must be devised to suit the *local* conditions, so that an efficient use of locally available materials is made.

Pile composting materials *

In order to increase the rate of fermentation of materials and raise the gas output, materials should be piled and composted before feeding into the pit — fibrous materials, especially straw, grass, weeds and maize stalks — must be thus treated because some of them have a waxy layer on the surface. Otherwise not only is it a hard and lengthy process for them to rot, but once fed into the pit, they float up to the surface and tend not to mix evenly with the other material. To pile and compost, cut the material into short pieces and pile up in layers, each layer about 50 cm thick. It is best to sprinkle on some material with a 2-5% lime or ash content, and then also pour on some human or animal manure or waste water, than cake over the surface with clay. In summer this piling and composting should last seven-10 days and in winter one month. When material has been thus handled, the waxy surface layer is broken down, which in turn hastens the breakdown of the fibrous materials in the material. Moreover, cutting the plant material increases the surface area of the material where it associates with the micro-organisms, which speeds up the fermentation process in the pit. When human or animal manure has been rotted in this way and then fed into the pit, it takes in with it natural organisms that produce biogas which then breed very quickly in the pit and greatly augment the rate of gas production. Also, according to studies, the carbon/nitrogen ratio for old and ripened vegetable material is 60:1-100:1. But with pile rotting, its carbon/nitrogen ratio can be reduced to 16:1-21:1, which approaches the ideal environment for the methane microbe.

Table 6-3. Water content of common fermentation materials.

Type of Material:	Human manure	Solid pig manure	Liquid pig manure	Horse manure	Common wind-dried manure	Dried rice stalks	Cow manure
Water content: %	80	82	96	76	30-40	10-20	83

*It is very common practice in China to ferment all sorts of manures to improve them. They pile them up in long ridges and cake them over with clay. That is called 'pile and rot'. In the warm climate of south east China, however, this has not been found necessary (see Appendix II).

Suitable water content

The normal activity of the methane microbe requires about 90% water in the fermentation material and 8-10% roughage or solids. The water content of common fermentation materials is shown in Table 6-3.

In actual fact, in many places the amount of water put in the pit is calculated from the estimated water content of the fermentation materials. Usually water makes up 50% of the material in the pit. Where it is difficult to estimate the water content of the fermentation material, it is safer to make it too diluted rather than too concentrated. In winter the material may be more concentrated but in summer it should be more dilute.

Inletting material

Upon completion of the pit, the first filling should be plentiful. First put in the pile-rotted stalks, grass and weeds. Then put in human and animal manure through both the inlet and outlet compartments. Cow and horse manure should best be mixed with water first, and then poured in through the inlet compartment. Lastly, fill up completely with ordinary water in order to expel all gas from within the pit. Afterwards remove 20-30 50-litre buckets of water to allow room for the accumulation of gas. Alternatively, do not completely fill the pit, allowing volume for the storage of gas. With this method, however, during the first few days the gas stored will be very impure and should be let off a few times.

During filling, the gas hose should first be disconnected from the gas outlet pipe and where there are removable covers, or safety valves, these should be opened so as to avoid build-up in the pit during filling. When the filling is complete, reseal the cover and safety valve with clay.

After one or two days in summer, or one week in winter, the pit will begin to produce gas that can be used. With some pits, in the early stages of gas production, though the quantity of gas produced is quite large for ten days or more, or even longer periods, the gas cannot be lit. This is because the gas contains little methane and much carbon dioxide and other gases as a result of incomplete fermentation and consequently low production of methane; or if the fermentation material is too acidic, the growth of the methane-producing microbes is inhibited. In this situation nothing special need be done; one can just let off gas a few times and after a few days normal production will gradually resume. If after a few days the material is still acidic, put in a small amount of lime or ash to adjust the acidity (the pH).

Rational supply of material

About a fortnight after initially filling the pit, start adding new material, because some of the existing material has already been broken down in the process of producing gas, through the fermenting action of microbes. In order to maintain a plentiful supply of nourishment for the microbes, to ensure continual production of gas, one must keep supplying fresh material for fermentation, as well as water, and some of the old material should be replaced. A schedule should be established for fairly frequent replenishment and removal of material. At fortnightly intervals in summer and every few days in winter remove fermented material from the exit compartment and replenish fresh material through the inlet compartment. This should be a mixture containing not only stalks and vegetable matter but a suitable quantity of human and animal manure. If the biogas pit is linked to a toilet or pigsty, the human and animal manure flows naturally into the pit, making it only necessary to supplement this with pile-rotted vegetable matter and water. In changing the material one should first outlet, then inlet material. Do not extract so much material and manure that the openings of the inlet and outlet compartments are exposed, as this will allow gas to escape. The quantity of material removed should equal the amount put in. If there is a temporary lack of material for fermentation, one should simply put in enough water to maintain the previous water level and the same capacity in the gas tank, in order to retain pressure in the gas tank.

Stir frequently (see Figure 6-1)
By stirring the liquid frequently one can ensure contact between the methane-producing microbes and the fermentation materials, which will produce maximum gas. When pits are not stirred, the fermentation material settles into three layers. The top layer is scum with a high content of fresh material, very few microbes in both number and variety, and much acid is produced here. The middle layer is clear fermented material containing very little solids and also few microbes. The bottom layer is sediment and residue rich in many kinds of microbes, but low on fresh material; because it is under a high hydrostatic pressure, the gas produced is dissolved in the fermenting liquid and is not easily released. This makes it impossible to achieve a high production of gas.

The sediments can be brought up from the bottom by stirring; the concentration and the temperature in the pit correspondingly become more even, the methane-producing microbes reproduce more quickly, and greater contact between the microbes and the fermentation materials hastens fermentation of the materials. Stirring can also break up or prevent the formation of any scum on the surface. In the fermentation of all organic materials, the bubbles produced often contain minute particles which rise to the surface of the liquid; after a long time a thick layer of scum forms and prevents the biogas from rising up into the gas tank and also reduces its production. Stirring will break up and make the small bubbles combine and escape more quickly from the liquid into the gas compartment.

Method of stirring
In most rural, small-scale biogas pits one can use a large pole or other tool and poke into the fermentation compartment through either the inlet or outlet compartment and move it around (Figure 6-1). Or some of the liquid could be extracted from the outlet compartment and then

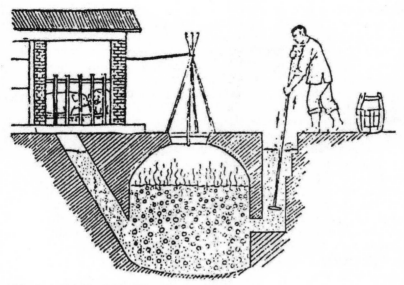

Fig. 6-1. Stir liquid fertilizer frequently.

poured back in through the inlet compartment, causing convection currents in the liquid in the pit. Sometimes a fixed frame is used to hold down in the fermenting liquid any rising impurities, scum or particles. Alternatively, use one or two bamboo or cane cylinders set vertically in the pit. This will reduce or prevent the formation of scum. But where such contraptions are fitted these pits should still be stirred. In large pits a mechanical device is needed for stirring.

Maintaining a suitable pH

Maintaining a suitable pH is also an effective means of raising gas production. Ideally, the pH in a pit should be a little on the alkaline side of neutral, with a pH 7.0-8.5. In order to maintain the necessary pH environment over the entire period of fermentation and during the inletting of material as well, one should frequently check and adjust the pH of the liquid.

The method of checking is simple. (1) Dip a piece of litmus paper into some of the fermentation liquid, immediately observe the change in colour, and compare this with a standard chart of colours to tell the pH of the liquid. (2) The people of Sichuan have observed that a red or yellow flame generally corresponds to slight over-acidity in the fermentation liquid.

If after a normal period of fermentation of the replenished material a slight acidity develops, or the gas produces reddish or yellowish flames when burning, one should remove some of the old material and replace it with a compensating amount of new material, or add some lime or ash, to adjust the acidity and restore normal gas production.

Measures to be taken in winter

In the winter the temperature within the pit falls and affects the production of gas to a certain extent; but as long as effective measures are taken in scientific management, it is entirely possible to maintain normal gas production and guarantee supplies. For example, in the 7th brigade of Donghe Commune in Zizhong County of Sichuan Province where a family of five had a rectangular stone biogas pit of 8 cu.m, they first filled it with material in December 1973, in total 2400 kg of human and animal manure, 200 kg of silkworm manure, 1000 kg of ginger leaves and stalks, and 400 kg of water. Throughout the winter they cooked three meals a day with the gas, and at night lit one gas lamp. When the outside temperature was $-1°C$, the temperature in the pit was $12\text{-}16°C$, and the pH was 7.0. There was normal gas production throughout the winter. In the biogas pit on exhibition at the All China Farm and Forestry Exhibition held at Peking, the pits were covered with earth, grass or straw during winter. In the coldest periods 100 kg of human or animal manure was added every 10 days; while the outside air temperature was $-14°C$, the temperature within the pit was $+15°C$. And the differential in the water column of the manometer was maintained between 130 cm and 140 cm.

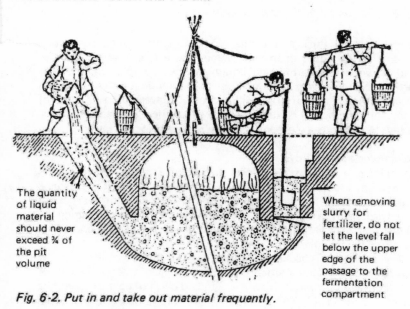

The quantity of liquid material should never exceed ¾ of the pit volume

When removing slurry for fertilizer, do not let the level fall below the upper edge of the passage to the fermentation compartment

Fig. 6-2. Put in and take out material frequently.

1. Take extra care to make frequent inletting and outletting of material (see Figure 6-2). This is one of the chief means of winter pit management. Before winter, the pit should be thoroughly changed once and filled with plenty of new material in order to ensure the necessary nourishment for fermentation during winter and to stimulate the methane-producing microbes to produce a

lot of gas. When removing the old material, leave one-third of the sediment layer. Better results will be obtained if the new material has been pile-rotted in advance. The fermentation material inletted should contain a little extra human or animal manure. The total quantity of material should be one-third more than during summer and the water should be decreased correspondingly. If possible, introduce a certain quantity of horse, dog manure, house-pets or silkworm manure, or dry grass, stalks and leaves of ginger, in order to raise the temperature of the fermenting liquid. A fortnight after changing the material, keep replenishing new material with frequent inletting and outletting. It is best to put in and take out a little every day.

Plate 6-1. Piggery built with direct flow of animal wastes into biogas pit inlet.

When the biogas pit is linked to toilets or to pigsties (Plate 6-1), the natural flow of human and animal manure into the pit will guarantee a plentiful supply of nourishment, and is a great advantage in raising winter gas production.

2. Proper temperatures must be maintained in the pit. In winter the temperatures are rather low, so some means must be devised to keep the pit's fermentation temperature high to ensure normal micro-organism activity. Such measures include covering the biogas pit and the inlet and outlet compartments with earth, grass or pile-rotted material. And when introducing new material, try to put in a little more exothermic fermentation material.

   Biogas pits which are linked to toilets and pigsties only need a cover over the outlet compartment to keep constant temperature.

3. Frequent stirring. Practice has shown that frequent stirring of material in the pit is also effective in maintaining winter gas production. It is best to stir once or twice every day.

4. Maintaining a proper pH balance in the fermentation liquid is also very important during winter, and frequent checks should be made. Normally the conditions should be slightly more alkaline than neutral with a pH of 7-8.5.

   The production of gas in winter is also directly related to the choice of site for the pit and the quality of construction.

7. Using Biogas

The chief component of biogas is methane (CH_4). This is an important raw material in industry, and an excellent gaseous fuel. In rural areas it is used mainly for cooking, lighting and preparation of fodder; and where conditions allow it can be used to drive machines and pumps, or to generate electricity.

Normal combustion of biogas with plenty of oxygen produces carbon dioxide, water and a great amount of heat, as illustrated by the formula:

$$CH_4 + 2O_2 \xrightarrow{\text{combustion}} CO_2 + 2H_2O + \text{heat}$$

For unit volume of methane we need twice the volume of oxygen to ensure complete combustion under stoichiometric conditions. The oxygen content of air is roughly one-fifth, so complete combustion of a unit volume of methane requires ten unit volumes of air. Since biogas contains 60-70% methane, complete combustion of a unit volume of biogas requires six or seven unit volumes of air. Fulfilling these conditions will release the maximum amount of heat.

In using biogas as fuel you should try to achieve a ratio of biogas to air that will allow complete combustion, in order to yield the best results. During complete combustion of biogas the flame is forceful, pale blue in colour, and makes a hissing sound. If it wavers and is pale blue in colour, there is too little air (oxygen) and incomplete combustion. If the flames are short, yellow and unsteady, then there is insufficient biogas and too much air — these conditions produce low temperatures and bad results.

Apparatus for using biogas in cooking and lighting

The gas hose
This should be rubber or plastic with an internal diameter of 0.6-1.0 cm. Some people use lengths of bamboo tube coated with pig's blood plaster connected with pieces of rubber hose, which also gives good results.

The section of hose out of doors is exposed to the elements and should be protected by a casing of bamboo tubing or other material.

Y's or X's (two way or three way branches)
These hose junctions are normally made of glass, but can also be made of metal or plastic.

Valves

Valves used on hose are glass, plastic, or if these types of valve are not available, one can make valves by the following methods:

1. Seal one end of a plastic hose 10 cm long, internal diameter 0.6-1.0 cm by heating, to make it airtight. Slip this over the gas-jet mouthpiece when not using the gas.

2. Bend a length of plastic or rubber hose over itself several times and then tie this up with string to block off the gas. Untie the string to allow the gas to flow.

3. Fit a length of soft rubber hose into the end of the gas hose and clamp this section when necessary.

4. Connect the gas hose to one arm of a T and on the opposite arm of the T insert a stick, wooden or bamboo, with a piece of cotton rolled on to the end. The cotton ball should be slightly larger in diameter than the diameter of the tube and should be coated with vaseline. When moved beyond the junction of the T it will close. Short of the junction it will open the valve. A hose should lead from the trunk of the T to the appliance.

A manometer

This is a simple instrument used to measure the pressure in the pit and also to calculate the rough quantity of gas stored. With such an instrument fitted, damage to the pit due to excessive internal pressure may be avoided. (See Chapter 5 on Maintenance and Quality Appraisal of Biogas Pits.)

The mouthpiece

This is a jet device fitted to one end of the hose. As the gas enters the burning appliance through this mouthpiece, its quality greatly affects combustion. One end of the mouthpiece should be slightly smaller in diameter than the hose, to ensure a tight fit. At the other end of the mouthpiece there should be a small hole through which the gas is ejected. One can use an appliance simply by inserting the mouthpiece into it.

A mouthpiece may also be made from bamboo, or the body of a ball point pen or a piece of glass tube. The dimensions of the small hole in the mouthpiece should depend on the rate at which the appliance uses the gas. For a gas lamp, the hole need only be the size of a needle for cotton thread. For a stove ring the hole should have a diameter of 1 mm. Because the quantity of biogas may vary during use, it is best to have on hand several mouthpieces with various sized holes.

In rural Sichuan the most commonly used mouthpieces are made from bamboo. Here are three methods of making bamboo mouthpieces:

The simple straight mouthpiece: Choose a piece of bamboo 0.5-0.6 cm in diameter with a notch in it. The notch should be regular, not askew. In the middle of the membrane of the notch, make a suitable hole with a needle. Shape the other end to fit the hose (see Figure 7-1). This type of mouthpiece is easy to make. The admixture of air is adjusted by altering the position of the supporting clamp, moving it back and forth, or by the size of the hole.

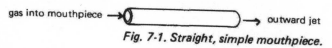

Fig. 7-1. Straight, simple mouthpiece.

Single-hole, controlled air mouthpiece: Choose a piece of bamboo with a notch 0.6-0.8 cm in diameter and 1.5-2.0 cm long on one side of the notch, and 4-6 cm on the other side. Shape the shorter end to fit the hose. Close to the notch on the longer section make a hole for the air inlet in one side, 1.3 cm long and 0.5 cm wide. In the membrane, on the opposite side from the air inlet hole, make a small hole (which in this case replaces the jet hole). The end of this longer section should be sharpened a little to make it easier to insert into the appliance (see Figure 7-2).

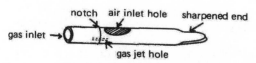

Fig. 7-2. Single-hole, controlled-air mouthpiece.

The advantages of this type of mouthpiece are that: the gas will not escape from the inlet hole; it is hardly affected by wind; once properly made it needs no adjustment by moving back and forth; and it is easy to make and use.

The three-pipe mouthpiece: This is made from a simple straight mouthpiece inserted into a piece of bamboo with four holes evenly spaced around it. Another sliding tube can move over the holes, and so adjust the amount of air let in (see Figure 7-3).

This type takes a little more trouble to make but makes less demands on the appliance itself, which then does not require a valve to control the air.

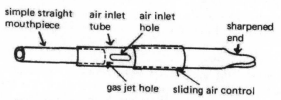

Fig. 7-3. The three pipe mouthpiece.

The gas stove
In various regions many types of stove are used, but the trend is towards the simple and convenient. We will here describe how to make a few of the more common stoves and burners.

i. Single tube burners
The bamboo tube burner: Take a piece of bamboo about 20 cm long and 0.8-1.2 cm in diameter, with notches at either end. In the membrane of one notch make a hole the size of the jet required (this takes the place of the jet mouth piece). In the membrane at the other end make a hole the size of a 'moong bean' (see Figure 7-4).

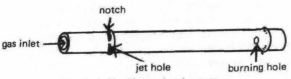

Fig. 7-4. Bamboo pipe burner.

Attach the hose to the end with the smaller hole and put the other end in the burner, pointing it straight at the bottom of the saucepan. The air admixture can be controlled by the distance between the top of the tube and the saucepan. This type of burner is extremely simple and easy to use.

Smoker's pipe burner: This is made from an old steel tube bent at one end, or can be moulded from any fire resistant material into the shape of a pipe. The diameter at the inlet end should be slightly smaller than the diameter at the burning end (see Figure 7-5).

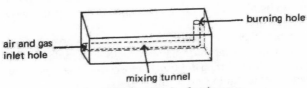

Fig. 7-5. Smoker's pipe burner.

The spiral single-pipe burner: This is made from old iron tubing 0.8-1.2 cm in diameter bent into a spiral. On one side of the spiral, along the arm, drill 30 holes 1 mm in diameter. This can also be formed out of clay.

The long arm burner: This appliance is an adaptation of the natural gas burner. Its advantages are that it gives a strong flame and is simple in construction. Iron and aluminium pots can both be used on this appliance. Construction as shown in Figure 7-7.

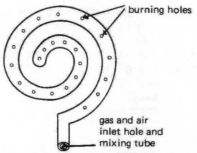

Fig. 7-6. The spiral single-pipe burner.

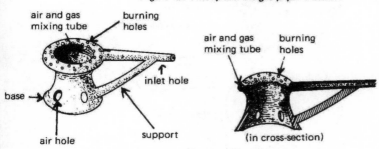

Fig. 7-7. The long arm burner.

ii. Showerhead-type burners

The showerhead burner: This type has a strong and concentrated fire power. Although normally made of potter's clay or other fire resistant material, made from cast iron it would be even longer lasting. The components are: base, gas inlet pipe, inlet-hole, gas and air mixing chamber and showerhead. The showerhead should be 7-8 cm in diameter and should have about 30 burning holes made in it. Diameter of base should be 12-15 cm and that of inlet pipe 0.6-0.8 cm. The air inlet hole should be 1 cm long and 0.5 cm wide (see Figure 7-8).

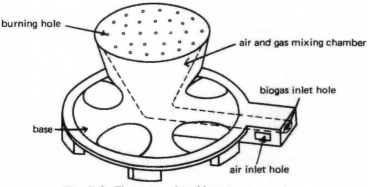

Fig. 7-8. The showerhead burner

The L-shaped burner: Basically made in the same way as the shower-head burner, it consists of a base, showerhead and an L-shaped gas-inlet pipe. On the head, drill holes 1.5-2 mm wide in three or four rows. This burner is controlled by a three-pipe mouthpiece. Its advantages are that the flames are strong and short, it economises on gas and functions readily (see Figure 7-9).

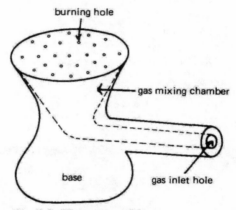

Fig. 7-9. The L-shaped burner.

The hourglass-shaped burner: (see Figure 7-10). This can be made simply by mounting a steel showerhead 5-7 cm in diameter on to a base made of triple concrete.

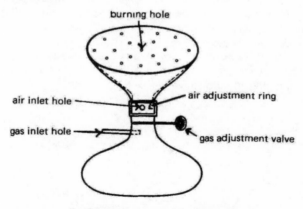

Fig. 7-10. Hourglass-shaped burner.

iii. Sieve-type burners

The drum burner: Here use whatever material is easily available to make a cylindrical drum 15-20 cm in diameter. The inside should be empty since this forms the gas/air mixing chamber. On the top drill 30 or 40 burning holes. It is best to drill them from the inside, outwards. Cover the bottom. At the lower end of the drum drill a gas inlet hole 0.6-0.8 cm in diameter (see Figure 7-11). One may also fit a tube 10-15 cm long into this hole (diameter 0.6-0.8 cm) and on the tube one can make a hole 1 cm long and 0.5 cm wide as the air inlet hole. This will make it even easier to use. Because the biogas and air can be well mixed in the drum mixing chamber before burning, combustion in this type of burner is very efficient.

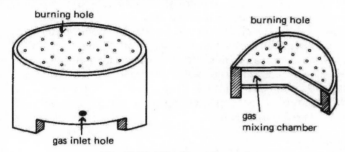

Fig. 7-11. The drum burner.

The revolving burner: This burner consists of a base and a cover. It has an air inlet hole at the centre, and an enclosed ring for the circulation of gas and air. To one side there is a gas and air inlet hole and in the cover are burning holes. During burning, gas is ejected from the holes, with a continual supply of air from the hole in the centre. There is a good supply of air so the flame is strong.

The construction method is to mould it from clay and fine cinder powder. First make a circular base, and in the centre make the air hole 2 cm in diameter. Around the air hole, along the top, make a circular groove 1 cm in diameter. Then according to the size of the burner make a cover in which there should be an air hole the same size as that in the base. In the region surrounding the air hole, drill a number of holes 1 mm in diameter, putting in three holes for every 2 cm. Then the cover should be fitted on the base and the interfaces sealed airtight.

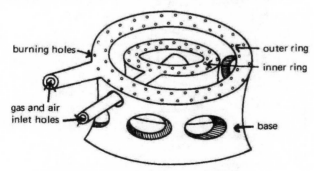

Fig. 7-12. The double ringed burner.

iv. Other types of burners
The double ringed burner: This type of burner is suitable for iron frying pans. The way it is made is similar to the method for the long handled burners, but the inner ring is slightly smaller than the outer ring and also a little lower (see Figure 7-12).

The biogas stove with cooking vessel actually built into the stove: This is a fairly recent development in the use of biogas. The gas is led directly into the cooker with the gas jet pointing directly at the bottom of the saucepan. The mouthpiece is inserted through a hole in the bottom of the stove cavity, which also lets in a suitable amount of air for mixing and burning within the stove cavity. Iron or aluminium pans may be used. It is crucial in this type of appliance to ensure a certain distance between the mouthpiece and the hole in the bottom of the stove to achieve a proper mixture of air and gas in the cooker and around the bottom of the saucepan during combustion. It should make a hissing noise (see Figure 7-13).

How to make a gas lamp
A gas lamp consists of a gas inlet hole, an air inlet hole, an air inlet adjustment valve, a mixing tube, a fire resistant clay head and a gauze mantle. At present in common rural use there are two types of lamp: a table or desk lamp which rests on a base, and a hanging lamp. They are normally moulded out of a mixture of clay and powdered cinders, or pottery clay when it is available, or waste material. They may be covered with a glass shade to protect the gauze mantle and increase the luminosity.

The hanging lamp: With ready-mixed clay, make a bell or funnel shaped lampshade and a body about 10 cm long. Down the centre make a round hole of diameter 0.6-0.8 cm with a chopstick or similar long, thin stick.

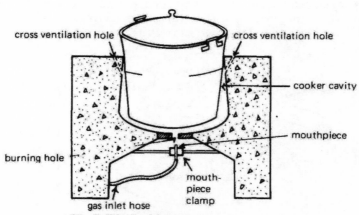

Fig. 7-13. The biogas stove.

One end of the shade should be enlarged so that the clay head can be fitted. At the other end, make four holes 1-1.5 cm long and 0.5 cm wide, evenly spaced round the top of the funnel shaped shade. If, however, a single-hole air-adjustment mouthpiece or a triple-mouthpiece is being used, these four holes are unnecessary. Next, hang it on a hook and, when dry, cover the head with the mantle and it is ready for use (see Figure 7-14).

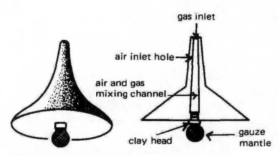

Fig. 7-14. The clay hanging lamp.

In some areas waste materials and iron plate are used to make the 'Red Star' hanging lamp (see Figure 7-15). This is durable and efficient but rather complex in construction.

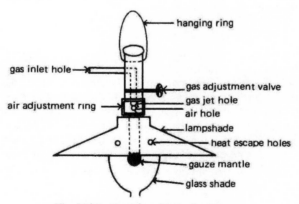

Fig. 7-15. 'Red Star' hanging lamp.

Table lamps: These are best fitted halfway up a wall and they burn very brightly. Because gas enters from below, the gas hose is not easily damaged. The gas and air tubes have a diameter of 1 cm and expand near the bends. Their size should accommodate a clay head. In certain areas a mobile lamp is also used, where the head is pointed upwards.

This type of lamp is quite efficient. The first time it is used it must be turned upside down — one must allow the mantle to set in its shape — and then it should be turned over and used.

Requirements for cooking and lighting with biogas

In cooking and lighting, the air supply has two routes (i) through the gas inlet hole; (ii) through air holes round the burning holes. Either of these two methods will facilitate speedy oxidation.

The suitable mixture of gas and air for complete combustion is related to the following conditions:

Pressure

When pressure in one arm of the manometer is more than 40 cm on the water column, the gas is forcefully ejected through the mouthpiece, such that it draws air into the inlet hole, and thus into the appliance. Thus by adjusting the valves, it is easy to achieve a suitable mixing ratio. But when the pressure in the pit is less than 20 cm of water, the force with which the gas comes out of the mouthpiece is small and it draws less air into the appliance. During use, especially if the mouthpiece is some distance away from the stove itself, the gas will have very little pressure and will not reach the appliance. If the mouthpiece is situated close to the actual light or stove, then the amount of air drawn in is insufficient and combustion will not be efficient. In practice we find that where the differential between the two columns in the manometer is in excess of 80 cm, we get good results, and where it is less than 40 cm, performance is poor.

The appliance

The quality of the appliance used for cooking or lighting bears directly on the efficiency of gas utilisation. A good appliance should have the following characteristics: (i) the inlet channels should be smooth to reduce the resistance to flow of gas and air; (ii) spacing and size of airholes should be suitable; (iii) the volume in the channel where the gas and air mix together should be large enough to allow complete mixture; (iv) the gas jet holes should not be too large but should allow easy passage of the mixed gas and air, to allow complete combustion; (v) the appliance should be simple, economical, strong, and cheap to make.

Insulation of ovens and stoves

Much of the heat produced from the burning of biogas is lost through convection and conduction. Therefore it is best to build stoves out of materials with low heat conductivity which, of course, are also fire resistant. More effective use of the heat may be obtained by placing the ring, or jet, inside a stove of some sort. The insulation may also prevent any incompletely combusted gas/air mixture from escaping by keeping it in the stove cavity to burn more completely.

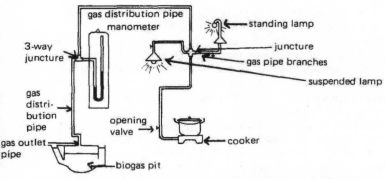

Fig. 7-16. Rough plan of equipment installation for using biogas.

Installation and Use

Before fitting, all hoses and valves should be checked for airtightness, and a rough plan (Figure 7-16) should be made of the positions where all valves, lights, stoves, and any Y or X joints should be fixed. Valves should best be fitted on to a wall (Figure 7-17). Next, cut the hose according to requirements and test it to make sure it fits. Reduce the number of connections and joins to a minimum. Connections between valves, hoses, branches and hose etc. should all be airtight.

Fig. 7-17. Wall attachment of biogas distribution pipes.

Use of stoves or heating appliances

When the stove is lit the flame should be pale blue, strong and even, and there should be a hissing sound. If the flame shifts around unsteadily, this may be because the mouthpiece has been inserted too far in the gas outlet hole and insufficient air has been drawn in. If after adjustment the flame still floats around, the quantity of air let in should be adjusted, or the small hole in the mouthpiece should be made smaller. If the flames have a red or yellow tint, there is too much

109

air or too little gas, and the air let in should be reduced, or the gas feed increased by enlarging the small hole in the mouthpiece.

When the ratio of air to gas is about 1:7, there may be a crackling noise after lighting. In this case the saucepan should be placed in position before lighting the gas, and it should be turned down low. It should only be turned up high after you have retracted your hand and sleeve, otherwise you will be burned. Turning the gas high too soon may also make it difficult to light.

Before using any appliance you should check it. Use a small tube to blow cigarette smoke through the mouthpiece and other holes to verify that they are not blocked. This will also enable you to check that the gas is distributing well as it comes out of the burning holes.

Use of lighting appliances

After a gas lamp has been built, first use it without the mantle and see whether the flame is pointed and pale blue in colour, and also if the gas comes out evenly from the little hole in the fire resistant head, making a hissing noise and not burning away from the fire resistant head. If these conditions are all satisfactorily met, the light is good. Put the mantle over the fire resistant head and light it, with the mouthpiece in the same position that was found to be suitable. The first time it is lit the gas pressure must be high. When the mantle is about to light up there will be one or two whishing noises, like the sound of something bursting into flame, as when a gas water heater lights up, or you set light to a small pile of gunpowder, and then the light will become bright. If the flame wanders, insufficient air is the cause. The air inlet should be adjusted, or the mouthpiece raised a little. If the flame still wanders, the hole in the mouthpiece is too large and should be changed for one with a smaller hole. If the flame becomes reddish, it indicates too little gas or too much air; the mouthpiece should be put further into the lamp to reduce the quantity of air going in. If this fails, the jet hole should be made larger.

When a pit first begins to produce gas the carbon dioxide and other impurities contained are rather high and the flames will tend to be reddish. This will cease upon normal production of gas. Some lamps do not produce incandescent light or white light, because the mantle is of poor quality, and a better one should be fitted. Lamps should be lit with a match or lighter with the flame held above the pouch of the gauze mantle. Do not light from below — this may cause carbon to darken the mantle and reduce the luminosity. If this happens, light the lamp, adjust the air inlet valve to shut off the air supply (or block the valve with fingers) and then open quickly. Do this repeatedly so that there is sporadic supply of oxygen. There will be more whishing noises and the carbon will gradually disappear.

8. Safety Measures in Building Biogas Pits and Using Biogas

In building the pit and using the gas, one must pay attention to safety, and widely disseminate knowledge about the safe use of gas. There should be frequent education or lectures on safety, and the necessary regulations must be established. Any slackness or slovenliness in this sphere must be overcome in order to avoid accidents.

Avoiding Construction Accidents

1. Preventing the pit collapsing or caving in during construction. If it is being built in the rainy season, drainage ditches should be built all around to avoid the accumulation of rainwater inside, because if the walls become soaked, the sides will cave in. A slope in the walls must be maintained during digging, and making overhanging walls should be strictly forbidden. Where the soil is loose it should be propped up and frequently checked.

2. In mining, transporting and building with stone, safety regulations must be respected and accidents due to falling rock avoided. When building a circular pit of stone slabs, they must be propped up with pieces of wood. Where arches have been made from round stones or triple concrete, remove the arch props with extreme care for safety. In brick arches built without arch props, the base brick must be firm. When mining stone, on all accounts avoid forming overhanging pieces of rock. When transporting stone, tie it on tightly. When pits are being made in sheer rock, or gravel, the site must be carefully chosen to avoid any sinking or caving in. During construction, when electric lighting is used, take great care — due to the dampness in the pit one must strictly forbid the use of faulty electrical appliances or cable. Avoid all contact with electric wires and appliances; in particular, avoid touching them with damp hands.

Preventing Poisoning and Suffocation

Biogas normally contains 60% methane. Although this is not a poisonous gas, when its content in the air reaches 30%, it can have an anaesthetising effect. When the concentration is about 70% it can cause asphyxiation or death from lack of oxygen. Furthermore, organic materials in the pit can produce poisonous gases in the absence of air. Therefore, in order to avoid poisoning or suffocation one must:

1. Make sure to remove all gas when emptying the pit. The partition

walls should not be made too low, and having a removable cover or a safety cover is a precaution.

2. As soon as material has been put into the pit, fermentation will begin and gas will be produced. Therefore if more stuff is to be put in, it should be let in through the inlet compartment or through the removable cover. On no account should it be loaded in by people going into the pit.

3. If it is necessary for people to go into the pit to change material, clear out any sediments, check or make repairs, then the gas hose should first be detached from the gas outlet pipe, and if the pit has a removable cover or safety hole, it should be left open for several days; a bellows, winnowing machine, or some machine to raise wind should be used to force air into the pit, so that any traces of gas will be expelled; only when there is plenty of fresh air inside, should people go in, or else they could be poisoned or suffocated. Before anyone enters the pit it is best to make a check with an animal. How is this done? Take a frog, a chicken, a rabbit — any such animal — tie a piece of rope or string onto it, lower it into the pit (see Figure 8-1) and bring it up again after a

Fig. 8-1. Take care that there is no trace of gas left in the pit, by flushing it out with air (left above), and by checking for air purity with an animal; then be sure to fasten on a safety belt before entering (courtesy Mianyang Science & Technology Committee).

few minutes. If the animal seems to behave normally, then this indicates that there is plenty of air in the pit, and one may enter it to work. If anything unusual should happen to the animal or if in fact it faints or loses consciousness, this indicates a lack of air in the pit or remnants of gas that have not been expelled, in which case do not enter the pit but continue circulating air.

4. After air has been pumped into and circulated through the pit and animal tests have been made, workers going into the pit should breathe through a long piece of pipe or hose with one end attached somewhere outside the pit. If one feels dizzy while working inside the pit, or finds it hard to breathe or in any way uncomfortable, leave the pit immediately and rest in a place where there is good air circulation.

5. Do not on any account introduce 'oil cakes' (oil cakes are what remains when oil bearing plants have been through an oil press; they are used a lot in China as fertilizer or as feed for draught animals and pigs), or phosphate fertilizer into the pit. This includes powder from phosphate mines or chemical fertilizers containing calcium, magnesium and phosphate, e.g. calcium hyperphosphate or superphosphate. In total absence of air these materials can produce the extremely poisonous gas phosphine (PH_3), contact with which may easily be fatal.

6. When making repairs do not use coke, charcoal or wood fires inside the pit to hasten drying the walls, since this will not only interfere with the quality of the pit, but also may cause coal gas poisoning.

7. If a person faints through lack of oxygen or from poisoning while working in a pit, he or she should be carried out immediately and placed where air circulation is good. If the person has stopped breathing, artificial respiration must be carried out, as well as heart massage if necessary. The person should immediately be taken to hospital, or a doctor should be brought to the scene to treat him or her. Those sent down to the rescue should be secured by rope before going into the pit and they should breathe deeply before going in and then breathe through a tube with one end secured outside of the pit. They should react with speed, but must not panic, as this would only result in more poisoning.

8. When clearing out sediment from the pit do not touch directly with hands or feet, but wear rubber gloves and boots.

Preventing burns
Biogas is extremely flammable and will burn fiercely. So it is important to avoid fires and burns.

1. When entering the pit to let out material, to check for leaks, or make repairs, do not take in paraffin or oil lamps, candles or anything with a naked flame. Also do not smoke in the pit or near the entrance. For illumination inside the pit use an electric torch or flashlight. Alternatively, create illumination by reflection off a mirror from outside. When removing the cover, once again do not have a naked flame, match or lit cigarette nearby.

2. Make frequent checks for air leaks in the hose and valves. If any hose has suffered damage through rat bites or erosion, it must be replaced in order to prevent rooms filling up with biogas and fires or poisoning occurring. The appliances for lighting or heating should be placed away from fodder, fuel, firewood, clothes, mosquito nets, and other flammable materials; special care must be taken in thatched houses to make sure that the light is a good distance away from the ceiling or roof.

3. Air must be circulated for ventilation indoors. Should one notice a strong smell of rotten eggs (the smell of the H_2S in the biogas) on entering a kitchen or room, then one should immediately open doors and windows in order to expel the gas; on no account should one light cigarettes or smoke in the room, so as to avoid fires.

4. After using an appliance, the valve should be shut tight. In the event of fire, one must shut the valves, detach the gas outlet hose from the gas outlet pipe, or block it in any way that will cut off gas supply.

5. In using biogas one should light the match first, and then open the gas valve. If the valve is opened and gas allowed to flow without being lit for any length of time, large amounts will leak out, spread out and rise in the room, and any flame will lead to severe fires.

6. Children must be taught not to play with fire close to the inlet and outlet compartments, so as to avoid burns.

Preventing explosions
In a closed space where there is a mixture of gas and air, such that the gas content is 15%, explosions can easily occur. Preventative measures which may be taken are:

1. Where a newly built pit has been filled with material and one is checking to see whether or not gas is being produced, the gas should be channelled away to some appliance on which tests can be made. It is strictly forbidden to test by holding a light to the

end of the gas outlet pipe, since the fire could easily be drawn back into the pit where the sudden expansion of the gases would cause an explosion.

2. When material has just been removed from the pit, do not use the gas immediately. If you must use the appliance, before lighting carefully check the direction of the gas flow in the mouthpiece, by holding chicken feathers or the down of a bird near the front of the mouthpiece. From the motion of the feather one can determine the direction of gas flow. If air is being sucked in, then do not hold the flame to it, for fire may be sucked into the pit which may lead to explosions.

3. During water tests or when filling the pit with material, especially when filling up past the passages in the inlet and outlet partitions, reduce the rate of fill so as to avoid a sudden rise in gas pressure inside the pit which could cause cracks to form. When removing material, also take care not to do this too quickly as it could create a vacuum which could also damage the walls of the pit. When large quantities of material are being let into, or taken out of the pit, detach the hose from the gas outlet pipe and take the removable cover off, to avoid the build-up of excessive positive or negative pressures. Take care to do this especially when pumping water in or out of the pit.

4. Once a pit has begun full-scale gas production, make frequent use of the gas to avoid an excessively large build-up of gas inside, and excessive pressures which can damage the pit. It is best to fit a manometer to every pit.

5. Pressure tanks on top of pits should have their water frequently replenished so as to avoid cracks from exposure to sun, and subsequent gas leaks.

Preventing drowning
The liquid in the biogas pit is quite deep and as the gas storage and fermentation tank, which is connected both to the inlet and the outlet compartments is sealed tight, it is extremely dangerous if man or beast were to fall into the pit. Therefore, over the inlet and outlet compartments it is absolutely necessary to make a cover out of very solid material such as wooden planks, or stone slabs. Furthermore, children should be taught not to play near the pit, so as to avoid falling in.

Intensifying safety education and establishing a safety management scheme
In order to ensure safety during construction of the pit and use of the gas, and to prevent accidents and protect human life and property, it is important to intensify the safety education during the popularisation

of biogas Scientific knowledge of safety in pit construction and safe use of gas should be disseminated. Through mass discussions necessary safety regulations should be implemented, and checks should be made from time to time to make sure that they are in force. Any carelessness must be banished, as should any attitude making light of safety. When problems are discussed they should be tackled immediately and any accidents prevented before they happen.

Appendix I

MATERIALS OF CONSTRUCTION

To permit a rational and efficient use of local materials, to guarantee the quality of pits and to reduce costs, we provide some statistics on building materials.

Stone

Because natural stone is hard and very durable, it is a good building material and is used a lot in biogas pits.

Table 1.1 Properties of stone

	Elastic Moduli (kg/cm^2)			Specific Gravity	Expansion Coefficient ($\frac{cm}{cm}$ °C $\times 10^{-7}$)	Absorption Rate of Water %	Durability (Years)
	Bulk Modulus	Rigidity	Young's Modulus				
Granite	1100-2100	80-150	130-190	2.6-3.0	34-118	0.2-1.7	75-200
Limestone	280-1400	18-200	70-140	2.6-2.8	30-112	0.1-6.0	20-40
Marble	700-1100	60-160	70-120	2.5-2.9	68-92	0.1-0.8	40-100
Sandstone	500-1400	35-140	85-180	2.5-2.7	67-116	0.7-13.8	50-200
Stone slabs	—	490-780	—	2.7-2.9	63-88	0-1.3	

Triple Cement

Triple cement is made from lime, mixed with either sand, cinders or stone chips or unspecified shape (or small pebbles) and clay. Following the addition of a suitable quantity of water the mixture is mixed thoroughly, pounded and settled, and finally slapped to smooth the surface.

Table 1.2 Properties of Triple Cement

Grade	Ratio (by volume)		Bulk Modulus after 28 days kg/cm^2	Notes: Dimension of samples (cm)
	lime, sand, clay	lime, cinders, clay		
1	2:2:8		20.9	15 x 15 x 15
2	2:3:8		19.05	15 x 15 x 15
3	1:4:5		11.6	20 x 20 x 20
4		1:3:5 (Stone)		
5		1:4:6 (Clay)		Ratios by weight
6		1:10:0.5 (Cement)	14-16	
7		1:10:0.1 (Gypsum)	13-15	

Brick

Bricks are commonly made from clay; normal machine-made bricks have a grading of 100-200; hand-made bricks have gradings from 50-75.

Table 1.3 Grades and properties of bricks

Grade of Brick	Limiting value of modulus (kg/cm^2)			
	Bulk		Young's	
	Av'ge	Min	Av'ge	Min
200	200	150	34	17
150	150	100	28	14
100	100	75	22	11
75	75	50	18	9
50	50	35	16	8

Notes:
1. *'Minimum' value means the lowest value found in a sample of 5. 'Average' is the arithmetic mean of the modulus from 5 samples, with no more than 2 below the average.*
2. *Where the sample is not representative or a larger sample is needed, one may take 3 sets of 5 etc. and thus grade the bricks.*

Cement

Cement is a wet-setting material capable of adhesion and solidification. Its main components are limestone and clay, which when adequately mixed and burnt under high temperature will partially melt down: the main component of the product is $CaSiO_3$. Adding a certain amount of gypsum or mineral dregs and grinding it all into a powder will then produce cement. The cement currently produced in factories is mostly over Grade 400, and in commune collective units mostly Grade 200 or 300. The higher the grade of cement, the greater its resistance to compression and tension.

Cement must be carefully stored to guard against damage by moisture. In rural communities cement purchased in loose or broken packs is best preserved in a covered pottery container. If the cement has been moistened it will congeal into hard pieces, which will lower its original adhesive capacity and affect the quality of the concrete or mortar for which it is used. The simple and convenient way of examining the level of moisture damage is to test how well it hardens. Previously moistened cement can be used if treated as follows:

1. If there are no hard pieces but chunks which can be crushed into powder, then pulverize them, and be sure to stir the cement more when using it.

2. If part of it has congealed into hard pieces, sieve them out and crush them into powder.

3. If most of it has already hardened, it must be crushed into powder

(or fried in an iron-pan up to 60°C), mixed with fresh cement (the old cement must be no more than 25%).

Once treated by any of the above methods, test the cement to determine its new grade and use accordingly. Cement should not be stored too long — under ordinary conditions, three months storage will lower its strength by 20% (the longer the time, the lower the strength). Once the cement has been stored more than three months its strength must be tested before it can be used. Cements of different varieties and grades must be transported and stored separately and the first lot delivered should be the first used.

Table 1.4 Properties of cement
(Maxima and minima as function of age)

Cement Grade	Bulk Modulus (kg/cm^2)						Young's Modulus (kg/cm^2)					
	3 days		7 days		28 days		3 days		7 days		28 days	
200	—	—	100	90	200	200	—	—	12	11	18	18
300	—	—	180	140	300	300	—	—	15	14	22	22
400	160	—	260	190	400	400	15	—	19	18	24	24
500	220	—	350	270	500	500	19	—	23	22	27	27
600	260	—	420	—	600	—	21	—	27	—	32	—

Mortar

Mortar is a mixture of adhesive, fine solid particles and water. The adhesive is usually cement, lime or gypsum, and the fine solid is usually sand and sometimes stove cinders. According to the intended purpose there are two kinds of mortar: building mortar and plastering mortar.

Table 1.5 Building mortars
Ratios of cement:lime:sand (by weight) for different grades of mortar.

Cement Grade	Mortar Grade				
	100	75	50	25	10
600	1:0.4:4.5	1:1.7:6	1:1.2:9	1:2.1:15	
400	1:0.3:3	1:1.3:4	1:0.7:6	1:1.7:6	1:2.1:14
300	1:0.1:2.5	1:0.2:3	1:0.4:4.5	1:1.2:9	1:2.1:12
200			1:0.1:5	1:0.5:5	1:1.7:10

Table 1.6 Mortars and plasters
Ratios of cement:sand (by weight)

Cement Grade	Mortar Grade		
	100	75	50
500	1:4.3	1:4.8	1:5.8
400	1:3.8	1:4.3	1:5.5
300	1:3.5	1:4	1:5

Table 1.7 Other mortars

Mortar type	Cementing material	Ratio (volume)	Grade
Clay	Clay	Clay:Sand = 1:3-5	3
Lime	Limewater	Limewater:Sand = 1:3-6	2-4
Lime	Unslaked Lime	Lime:Sand = 1:3-7	4-10
Lime & Clay	Unslaked Lime	Lime:Clay:Sand = 1:0.3:5-7	2-4
Plaster of Paris	Industrial Plaster	Plaster:Sand = 1:1	50
	of Paris	Plaster:Sand = 1:1.5	25
	(for construction)	Plaster:Sand = 1:1.3	10

Common wet setting mortars have grades 10, 10, 50; the water used should be clean and should be used sparingly — too much will impair the quality. Actual quantities to be used for various mortars are given in Table 1.8.

Table 1.8 Wet-setting mortars

Material	Unit	Grade				
		10	25	50	75	100
400 Cement	kg	85	125	184	260	320
Unslaked lime	kg	110	100	56.6	45.5	40
Sand	m³	1.01	1.01	1.01	1.01	1.01
Water	m³	0.4	0.4	0.4	0.4	0.4

Table 1:9 Wet-setting mortars
Using 300 grade cement, mortars of grades 25 and 50 may be made.

Material	Unit	Grade of Mortar	
		25	50
Cement	kg	137	235
Unslaked lime	kg	85.3	60.3
Sand	m³	1.013	1.015
Water	m³	0.4	0.4

(to save cement it is possible to use other mortars)

Table 1.10 Coating mortars

Material	Ratio	Use	Note
Lime:Sand	1:2-1:5	On brick surfaces	Non-drip
Cement:Lime:Sand	1:1:4-1:1:6	On Wet Regions	Consistency
Cement:Sand	1:2-1:3	On Wet Regions	

Table 1.11 Formulae for calculation of areas

	AREA
L (square with sides L and L)	$L \times L$
H, L (rectangle)	$H \times L$
M, H, L (trapezium)	$\dfrac{(L + M)}{2} \times H$
H, L (parallelogram)	$L \times H$
H, L (triangle)	$\dfrac{L \times H}{2}$
R (circle)	πR^2 $= 3.142\,R^2$

Table 1.12 (Volumes)

	VOLUME
	$L \times L \times L$
	$L \times H \times W$
	$\dfrac{(A_1 + A_2)}{2} \times H$
	$A_1 \times H$
	$\approx \dfrac{(2\pi R)^2 \times .08 \times H}{3}$

Conversion Tables

Table 1.13 Lengths

m	cm	British feet	Chinese feet	Chinese inches
1	100	3.281	3	30
0.01	1	0.0328	0.03	0.03
0.305	30.5	1	0.914	9.14
0.333	33.333	1.094	1	10

Table 1.14 Weights

Tonnes	kg		gms	lbs
1	1,000	2,000	1,000,000	2204.6
0.001	1	2	1,000	2.2046
0.0005	0.5	1	500	1.1023
0.00045	0.4539	0.9078	435.90	1
0.000001	0.001	0.002	1	0.0022

Table 1.15 Areas

m^2	ft^2	(Chinese ft^2)	cm^2
1	10.7650	9	10,000
0.0930	1	0.8353	930.25
0.1009	1.1689	1	1008.89

Table 1.16 Volumes

m^3	ft^3	Gallons (water)	Litres (water)
1	35.3165	219.969	1,000
0.02832	1	6.2355	28.3168
0.00455	0.1604	1	4.546
0.001	0.03532	0.21997	1

Table 1.17 Pressure

Atmos.	Kg/cm^2	p.s.i.	mm Hg	mm H_2O
1	1.0332	14.696	760	10,323
0.9678	1	14.223	735.56	10,000
0.0680	0.0703	1	51.715	703
0.00131	0.00136	0.0193	1	13.6
0.000097	0.0001	0.00142	0.07356	1

Table 1.18 Some properties of gases

Name	Symbol	Mol. Wt.	Density kg/m^3 at S.T.P.	Spec* Gravity	Critical Temp. ($^\circ$C)	Critical Pressure (atm)	Volume on liquefaction of $1m^3$ (litres)
Hydrogen	H_2	2.016	0.0899	0.0695	-239.9	12.8	1.166
Carbon Monoxide	CO	28.01	1.2504	0.9669	-140.2	34.54	1.411
Carbon Dioxide	CO_2	44.01	1.977	1.529	$+ 31.1$	73.0	1.56
Oxygen	O_2	32.0	1.429	1.105	-118.8	49.71	1.15
Methane	CH_4	16.04	0.716	0.554	$- 82.1$	45.8	1.55
Ethane	C_2H_6	30.07	1.357	1.049	$+ 32$	48.2	2.25
Air	–	28.95	1.293	1	-140.7	37.2	1.379
Water (vapour)	H_2O	18.02	0.805	0.594	$+374$	224.7	0.737
Chlorine	Cl_2	70.91	3.214	2.486	$+144$	76.1	2.006
Acetylene	C_2H_2	26.036	1.173	0.9057	$+ 36$	61.7	2.055

*with respect to air

Table 1.19 Explosive Limits

Name			Explosive Limits (% by volume)			
			in air		in oxygen	
			lower limit	upper limit	lower limit	upper limit
Hydrogen			4.1	75	4.5	95
Carbon Monoxide			12.5	75	13	95
Methane			5	15	5	60
Ethane			–	15	4	50
Acetylene			1.95	8.2	2.8	93
Hydrogen Sulphide			4.5	45.5	–	–
Ammonia			14	33	12.6	80

Appendix II

First-hand observations

Edited transcript of an interview with Liang Daming, Head of the Biogas works Office, Shachiao People's Commune, Shunde County, Guangdong Province. 28 September 1978.

Shachiao Commune: Approx. 40 km southeast of Guangzhou (Canton), downstream on the Po River.
Population: 72,000. 25 production brigades subdivided into 311 production teams.
Arable land: 4,433 hectares.
Economic activities: fish farming, sugar cane, mulberry leaves (to feed:), silk worms, some cash crops. Primary export: 15 tonnes of carp per day to Canton.

In 1976 Guangdong Province sent three people from Shachiao Commune to Sichuan Province to acquire and bring back the expertise necessary to set up their own biogas programme. This was the second group to be sent from Guangdong to Sichuan, to benefit from the experimentation with biogas which began there in 1958 following a directive delivered by Mao Zedong in Wuhan. He called on the people in the countryside to make optimum use of local materials and to begin to develop biogas technology as part of the Great Leap Forward and the establishment of collective rural organisation.

Of the three people from Shachiao, only one had experience in building. Liang Daming and the third were employed in the commercial department in charge of distribution of goods in Shachiao Commune. None of them had had any previous training in biogas technology.

They read this manual on the train to Sichuan — it was the only book they used — and then spent 15 days with families who were building and operating biogas pits.

On return, they immediately built the first pit in Shachiao. It took over one month to build and was a success. They then built four more, three of which succeeded and one failed. Unlike the mountainous terrain of Sichuan, Guangdong lies very close to sea level. The high water table forced them to abandon the use of 'triple concrete', because the lime which acts as the bonding material for sand and mud in the traditional mixture was causing their pits to leak. Instead they had to substitute grade 400 commercial cement, despite their preference for the cheaper lime.

When the construction technique had been perfected, demonstration pits were built. The building team invited the leaders of the commune, brigades and teams to observe them in operation. They took great pains to explain the benefits of biogas development. After the demonstration, these leaders took every opportunity to relay the message, in order to overcome peoples' suspicion that the biogas and slurry could cause disease (Plate II-1). Whenever they had a meeting they would say a

Plate II-1. Rural health station poster showing advantages of biogas and other forms of waste fermentation and composting.

few words about it. They drew big charts, made plans in the teams and brigades, paid visits to the homes of families to explain to them why they should do such a thing, and gradually people changed their outlook and resolved to give it a try. The first phase was therefore to propagate the ideas.

A new technique of pit construction was invented, with the bottom part of the pit pre-cast in concrete in a hollow carved to the right shape out of the soil at ground level *above* the water table (Plates II-2 and II-3).

Plate II-2. 20 cu.m collective biogas pit under construction between pigsties and fishpond; 'floating construction method'.

Right beside this concave concrete dish, a hole is dug for the pit, filled with water until the concave dish beside it floats and can then be floated over until positioned above the hole. The water is then pumped out, bringing the dish to rest in place at the bottom of the hole. A cork is then removed from the centre of the dish, so that it will not float upwards as the water seeps back into the hole. With the concrete dish as a firm base to stand on, workers can then proceed to build up the brick walls of the pit along the rim of the dish. Heavy clay is packed over the top of the completed fermentation tank to increase the downward pressure on the pit. Enough water is also kept in the pit at all times to prevent floatation.

127

Plate II-3. 20 cu.m collective biogas pit under construction; 'floating construction method': shallow mould for precast concrete pit shown at top; note water table risen inside and outside pit.

A team of 70 technicians was trained, among whom only a few were builders by trade. Each person had to learn the whole process and everyone to build with their own hands, so that after a year they had all become builders and experts. The seventy were chosen according to two criteria: they had to have a high sense of responsibility for the brigade and team; and at the end of their training, they *had* to be qualified, which meant that they had to work hard, while keeping costs to a minimum and even trying to reduce them. They were encouraged to put their heads together, to come up with their own ideas, and to take advantage of all new ideas.

Under the supervision of these technicians and the original experimenters, the brigades and teams undertook an active construction programme. They assembled the materials and people to build a pit in each brigade, so that these could acquire the experience through doing it themselves.

Experiments which had first been organised at commune level thus spread to brigade level. Each brigade leader was asked to build his own pit, assisted by teachers from the brigades' primary and secondary schools — this was one of the most important steps in popularising the technology. On that basis it was spread throughout the commune, down to the individual household.

By the end of 1976, several months after the programme's inception, 100 pits had been built by the brigades. By the end of 1977, another 600 were in operation. By September, 1978, just two years from when they first began learning about biogas, there were 1,640 pits in the commune, of which 1,300 were built by individual families. (This impressive achievement, which Liang called 'gradual', was attributed to their rapid adaptation of the construction principles to their own circumstances. Liang remained concerned, however, about the one pit in the initial experimental phase, which they still have not been able to make work — Ed.)

These pits range in size from three cu.m to 64 cu.m. The individual family pits are five to seven cu.m; the collective pits 30-40 cu.m. The former are used to produce fuel for cooking and lighting and for fertilizer; the latter are used to generate electricity for irrigation and pumping, for grinding pig fodder, and for lighting and the silkworm houses, with fertilizer, once again, as the other important product.

The three tasks set for the Biogas Office at its establishment were to:
— raise the efficacy of organic fertilizer
— transform the rural environment and increase the level of sanitation, and
— solve the fuel shortage by developing an inexhaustible source of energy.

The benefits of biogas have been the reduction of the peasants' burden by releasing them from the dependence on coal and firewood, a reduction in the amount of electricity the State must supply, and a reduction in the brigades' production costs since they now have to buy less fertilizer and less fuel for irrigation. They say that the fertilizer fermented in pits is improved in quality because less ammonia evaporates, and that it is better than chemical fertilizer because it keeps the soil crumbly and provides a steadier source of nutriment for the crops. Their acquisition of commercial fertilizer has been decreased by 30%.

Four types of pit were visited. Beside the silkworm houses there were several recently completed pits, each with a capacity of ten cu.m, a spherical fermentation compartment, a water tank on top, and all parts made out of concrete (Plate II-4).

Plate II-4. 10 cu.m spherical concrete biogas pit with water pressure tanks built on top, outlet compartments on left, inlet on right, removable cover sitting on the ground behind the pit.

The second was a collective pit of the largest variety, 64 cubic metres and rectangular (Plate II-5). It is used primarily to generate electricity (Plate II-6) for irrigation, for grinding pig fodder, and for the lighting requirements of 180 families. Its innovation is a low-pressure system of gas storage, where, when the gas pressure rises above $5°$ on the pressure gauge, gas flows over into two polythene storage balloons of 18 cu.m each (Plates II-7 and II-8). The third type was a spherical pit being built by the 'floating construction method' described above (Plates II-1 and II-2) out of concrete and bricks, with a planned capacity of 20 cu.m. Its location right next to a row of pigsties permits direct manure disposal (see Plate 4-2 in text).

The last examples were two family pits in operation. There was one underground and another, a rectangular concrete pit, was mostly above ground. The one underground supplied the needs of the family of the production team leader (Plate II-9), seven people in all. The pit had a volume of 7.3 cubic metres, and had been built with bricks and cement valued at 45 ¥uan and supplied to the family by the commune. The family had contributed the necessary labour time, 35 person-

Plate II-5. Inlet compartments to a 64 cu.m collective biogas pit for electricity generation; rectangular body of pit extends to the right, power station in background.

days. They were assisted by the commune's Biogas Group in both construction and maintenance. The family was responsible for collecting dung to feed into the pit (which they said took no longer than the time to smoke a cigarette!) for feeding it in, and for letting out gas according to the pressure gauge in their kitchen (Plate II-10).

Plate II-6. Electricity generating station interior with diesel generator running on biogas from the 64 cu.m collective pit.

Plate II-7. 18 cu.m polythene biogas storage balloons inside electricity generating station; two balloons store the gas from the 64 cu.m pit beside the station; Liang Daming, Head of Biogas Works Office in foreground.

Plate II-8. Polythene biogas storage balloon under repair on a city street (Guangzhou; courtesy, Rosemary Robinson).

Maintenance and operation

The 70 technicians are now in charge of safety and management, working full-time to check on maintenance, help with repairs, and generally be at the disposal of anyone having problems with a pit. In addition, they have passed this expertise on to another 300 people, who perform this service part-time.

Liang identified the following important points from their experience in maintenance and operation. The two decisive factors for proper gas production are the ambient temperature and the combination of materials fed into the pit. As a rule of thumb, the more organic the contents, the more gas is produced, and the more nitrogen, the better the fertilizer. One cubic metre of pit volume can produce 0.15-0.20 metres of gas per day at 25-33°C ambient temperature. Above 30°C the same volume can produce 0.3 cubic metres per day. On the average, the internal pit temperature is about 5°C higher, but this depends on the water level around the pit; water will absorb and draw off heat in sandy soil. At Shachiao the soil is dense clay and so heat loss is not a significant problem.

One should put material into the pit at least once every three days, but preferably every day. Each time, roughly 50 kg of fresh material are required for every cubic metre of gas required every day. Of this 20% should be organic and up to 80% can be water. Some paper can also go into the pit without hindering gas production but is not particularly advisable: paper contains little organic matter and mostly fibre — too large a quantity will simply build up, occupying valuable space inside the pit and thereby reducing the volume of gas produced.

At Shachiao they have had no shortage of materials to feed into their pits during winter but have simply found that when the ambient temperature goes down to 5°C too little gas is produced. Then they cover the pit with rice straw to insulate it, and feed the pit more frequently. They empty the pits once or twice a year to remove fibre and to conduct repairs. Otherwise material is continually put in and taken out.

Liang does not forsee that they will have many problems, except

Plate II-9. Biogas land (and alternative electric bulb) in the house of a production team leader (sitting lower right; beside him one of the commune's leaders).

that the plastic they use for the pipes and taps is expected to harden and will have to be replaced every three years or so. They have not found it necessary to stir the pit contents, as the convection currents from inletting and outletting cause sufficient mixing.

Shachiao exchanges information with other communes primarily for the purposes of updating their knowledge. However, they do not engage in any direct or active education about the technology.

Plate II-10. Biogas stove and manometer in kitchen of production team leader's family.